AF803713

Advances in Analytics for Learning and Teaching

Series Editors
Dirk Ifenthaler, Learning, Design and Technology, University of Mannheim, Mannheim, Germany
David Gibson, Teaching and Learning, Curtin University, Bentley, USA

This book series highlights the latest developments of analytics for learning and teaching as well as providing an arena for the further development of this rapidly developing field.

It provides insight into the emerging paradigms, frameworks, methods, and processes of managing change to better facilitate organizational transformation toward implementation of educational data mining and learning analytics. The series accepts monographs and edited volumes focusing on the above-mentioned scope, and covers a number of subjects. Titles in the series *Advances in Analytics for Learning and Teaching* look at education from K-12 through higher education, as well as vocational, business, and health education. The series also is interested in teaching, learning, and instructional design and organization as well as data analytics and technology adoption.

David C. Gibson • Dirk Ifenthaler

Computational Learning Theories

Models for Artificial Intelligence Promoting Learning Processes

Springer

David C. Gibson
UNESCO Chair on Data Science in Higher Education Learning and Teaching
Curtin University
Bentley, WA, Australia

Dirk Ifenthaler
Learning, Design and Technology
University of Mannheim
Mannheim, Germany

ISSN 2662-2122 ISSN 2662-2130 (electronic)
Advances in Analytics for Learning and Teaching
ISBN 978-3-031-65897-6 ISBN 978-3-031-65898-3 (eBook)
https://doi.org/10.1007/978-3-031-65898-3

© The Editor(s) (if applicable) and The Author(s), under exclusive license to Springer Nature Switzerland AG 2024
This work is subject to copyright. All rights are solely and exclusively licensed by the Publisher, whether the whole or part of the material is concerned, specifically the rights of translation, reprinting, reuse of illustrations, recitation, broadcasting, reproduction on microfilms or in any other physical way, and transmission or information storage and retrieval, electronic adaptation, computer software, or by similar or dissimilar methodology now known or hereafter developed.
The use of general descriptive names, registered names, trademarks, service marks, etc. in this publication does not imply, even in the absence of a specific statement, that such names are exempt from the relevant protective laws and regulations and therefore free for general use.
The publisher, the authors and the editors are safe to assume that the advice and information in this book are believed to be true and accurate at the date of publication. Neither the publisher nor the authors or the editors give a warranty, expressed or implied, with respect to the material contained herein or for any errors or omissions that may have been made. The publisher remains neutral with regard to jurisdictional claims in published maps and institutional affiliations.

This Springer imprint is published by the registered company Springer Nature Switzerland AG
The registered company address is: Gewerbestrasse 11, 6330 Cham, Switzerland

If disposing of this product, please recycle the paper.

Preface

This book proposes a groundbreaking multi-level framework for understanding learning. It integrates theories from psychology, education, and AI to explore how learning unfolds across individuals, teams, organizations, and cultures. A key concept is the **3-level network model** that captures the essence of learning processes. This model is used to design AI applications that can support learning at each level, ultimately fostering increased agency, self-awareness, and self-regulation. The book delves into the micro level (individual learning), the meso level (team and organizational learning), and the macro level (cultural learning). Each level is explored through dedicated chapters that introduce relevant theories and network models. The book also addresses the ethical considerations and practical applications of AI in enhancing learning across these different contexts. This framework offers a powerful tool for researchers and practitioners to understand learning as a complex and interconnected phenomenon. It paves the way for the development of next-generation AI tools that can empower individuals, teams, and cultures to learn and grow more effectively.

Organization of the Book

Chapter 1 Why 'Computational' Learning Theories?

Chapter 1 argues that to leverage AI for improved learning, learning theories must become computational models, incorporating complexity and machine learning for applications across all educational scales.

Chapter 2 AI and Learning Processes

This chapter sees AI as a tool to improve learning at all levels (individual, team, culture) by focusing on learning processes as the core driver of educational success, proposing a new framework that merges learning theories and AI for deeper understanding.

Chapter 3 A Complex Hierarchical Framework of Learning

Chapter 3 introduces a complex hierarchical network framework of learning, drawing from developmental psychology, computational biology, instructional design, cognitive science, complexity, and sociocultural theory.

Chapter 4 Piaget and the Ontogeny of Intelligence

This chapter explores how intelligence develops through a 4-phase cycle, applicable to individual, team, and cultural learning.

Chapter 5 Keller and the ARCS Model of Motivation

Chapter 5 connects Piaget's theory of equilibration with the ARCS model of motivation to explore how learning drives sustainability and growth.

Chapter 6 Complexity Theory and Learning

This chapter proposes a new model for individual learning, merging physics, psychology, and self-organizing systems to show how learning drives intelligence evolution.

Chapter 7 AI Roles for Enhancing Individual Learning

Chapter 7 leverages the micro-learning network model to design AI tools that support individual learning journeys.

Chapter 8 Informal Social Learning

This chapter establishes the meso level, emphasizing social interaction and real-world experiences in learning, and builds on micro-level theories to introduce a formal social cognition network.

Chapter 9 How People Learn

This chapter proposes a 4-node network for meso-level learning by teams and organizations and introduces AI's role in boosting their agency.

Chapter 10 AI Assisting Individuals as Team Members

Chapter 10 outlines meso-level AI development concepts that emphasize balancing individual needs with the team's mission while prioritizing the team's success and agency through learning.

Chapter 11 AI Roles for the Team or Organization

This chapter outlines AI roles for team success and argues that conflicts must serve the team's mission, not individuals, to avoid ethical issues and team dissolution.

Chapter 12 A Network Theory of Culture

Chapter 12 proposes a 3-level network model of cultural learning, with cultural historical activity theory at the macro level encompassing other theories and allowing for various analysis approaches.

Chapter 13 AI Roles in Cultural Learning

Building on a network theory of culture, this chapter explores AI applications to foster cultural exchange, understanding, and interdisciplinary collaboration.

Chapter 14 Open Questions

The final chapter critically reflects open questions arising from the previous chapters focusing on computational learning theories.

Key Features of the Book

This book shows how artificial intelligence grounded in learning theories can promote individual learning, team productivity, and multidisciplinary knowledge-building. It advances the learning sciences by integrating learning theory with computational biology and complexity, offering an updated mechanism of learning, which integrates previous theories, provides a basis for scaling from individuals to societies, and unifies models of psychology, sociology, and cultural studies.

The book provides a road map for the development of AI that addresses the central problems of learning theory in the age of artificial intelligence including:

- Optimizing human-machine collaboration
- Promoting individual learning
- Balancing personalization with privacy
- Dealing with biases and promoting fairness
- Explaining decisions and recommendations to build trust and accountability
- Continuously balancing and adapting to individual, team and organizational goals
- Generating and generalizing knowledge across fields and domains

The book will be of interest to educational professionals, researchers, and developers of educational technology that utilize artificial intelligence.

Bentley, WA, Australia David C. Gibson

Bentley, WA, Australia Dirk Ifenthaler
Mannheim, Germany

Contents

About the Authors

David C. Gibson is the UNESCO Chair on Data Science in Higher Education Learning and Teaching at Curtin University in Perth Australia. He received his doctorate in Leadership and Policy Studies from the University of Vermont in 1999 based on a study of complex systems modeling of educational change. His study demonstrated the feasibility of bridging from qualitative information to quantifiable dynamic relationships in complex models that verify trajectories of organizational change. He provides thought leadership as a researcher, professor, learning scientist and innovator. He is creator of simSchool, a classroom flight simulator for preparing educators, and he provided vision and sponsorship for Curtin University's Challenge, a mobile, game-based learning platform. His research has extended from learning analytics, complex systems analysis and modeling of education to application of complexity via games and simulations in teacher education, web applications and the future of learning. Dr. Gibson has also advanced the use of technology to personalize education via cognitive modeling, design and implementation.

Dirk Ifenthaler is Professor and Chair of Learning, Design and Technology at the University of Mannheim, Germany and UNESCO Deputy Chair on Data Science in Higher Education Learning and Teaching at Curtin University, Australia. Dirk's research focuses on the intersection of cognitive psychology, educational technology, data analytics, and organizational learning. He is the Editor-in-Chief of the *Technology, Knowledge and Learning* and Editor-in-Chief of *Educational Technology and Society*.

Chapter 1
Why 'Computational' Learning Theories?

1.1 Introduction

This book begins with the idea that learning theories must be *computational* theories if artificial intelligence (AI) is to fulfill its promise as an enhancement to human learning processes, thinking, and action. Why so? Fundamentally, because computation is the language of computers and thus artificial intelligence. This stance implies a corollary proposition: AI must use transparent and understandable *human* communication to fulfill its promise as an enhancement to human learning, as that is the language people understand.

The book introduces a 3-level complex system model (micro, meso, and macro) based on a definition of intelligence as a natural consequence of an agent's ongoing exploratory growth and learning processes driven by the urge to live (Kauffman, 2000). It holds that a ubiquitous feature of individual, group, and cultural agency is a causal learning cycle, which is the root cause of the drive to expand and increase the capabilities the agent needs to survive, reproduce, and evolve. The causal learning processes at micro, meso, and macro levels ultimately lead to cognition, knowledge, and creative capabilities at every level coevolving within the 'adjacent possible' of a local niche in the universe. With that context, the book outlines a framework for developing AI applications that can assist learning processes at all three levels.

The field of learning encompasses numerous theories that have evolved, offering diverse perspectives on how learning occurs. While it's challenging to provide an exact count due to variations in categorization and ongoing theoretical developments, an estimate suggests there are around 20–30 major learning theories in perhaps a dozen or more categories that have gained prominence in educational psychology and related disciplines (Zhou & Brown, 2015). However, this count can

© The Author(s), under exclusive license to Springer Nature Switzerland AG 2024

D. C. Gibson, D. Ifenthaler, *Computational Learning Theories*, Advances in Analytics for Learning and Teaching,
https://doi.org/10.1007/978-3-031-65898-3_1

vary based on the inclusion or exclusion of sub-theories, new emerging perspectives, or different categorization methods within the field and in terms of the unique contributions of fields like neuroscience, computer science and psychology.

1.2 Overlapping and Conflicting Ideas in Learning Theories

Among the reasons there are so many theories of learning is that each model explains but a part of the complex phenomenon. Each theory has arisen from a nuanced perspective that different models offer, each focusing on specific facets of the learning process, which has led to many overlapping ideas. For example, most theories include a common concern for active agency in learning; an agent's responses are active in a behavioral model, neural reorganization is involved in a cognitive model, and mental model elaboration takes place in a constructivist model.

In behaviorism, active participation is evident in the learner's responses to stimuli, shaping behavior through reinforcement or conditioning. For instance, operant conditioning explores how behaviors are shaped by consequences, promoting or inhibiting their recurrence. Cognitive theories emphasize active engagement, focusing on neural reorganization and brain plasticity as learners actively process and restructure information. For example, in cognitive load theory (Sweller et al., 2011), learners actively organize and reorganize information to facilitate learning. Constructivism highlights active participation through mental model elaboration. Learners construct knowledge by actively engaging with new information and assimilating it into existing mental frameworks or schemas. An example is the zone of proximal development (Vygotsky, 1978), where learning occurs in an adjacent space of possibilities through social interaction and active participation in problem-solving.

Situated learning theories (Brown et al., 1989) stress the importance of learning within authentic contexts, where knowledge is acquired through active engagement in real-world tasks or environments. An example is the Communities of Practice theory (Lave, 1991), where learning is embedded in social contexts through active participation in shared activities and discussions. Experiential learning theories emphasize learning by doing, where active participation in hands-on experiences facilitates knowledge acquisition. Kolb's Experiential Learning Theory for example (Kolb & Fry, 1975) illustrates this through the cycle of concrete experiences, reflective observation, abstract conceptualization, and active experimentation.

These examples demonstrate how despite differing perspectives and methodologies, various learning theories converge on certain shared principles. Active participation, context-based learning, and engagement with experiences emerge as common threads, offering insights into the multifaceted nature of learning and emphasizing the significance of learner involvement in the learning process across diverse theoretical frameworks.

However, the many theories have also led to conflicting ideas that often arise due to varying perspectives on the extent of agency or the capacity for self-directed action within the learning process. For example, behavioral theories limit agency to the learner's reactions and conditioned responses to environmental cues. Cognitive theories limit the scope of agency to the restructuring of mental representations and neural connections, ignoring the distribution of intelligence to others and the physical world. Constructivism likewise constrains agency to the boundaries of personal mental constructs.

Theories also diverge in their views of the expansiveness of learning agency. Humanistic theories, such as Maslow's hierarchy of needs (Maslow, 1943) or Rogers' person-centered approach (Rogers et al., 2015), for example, emphasize the learner's self-directed growth and potential for self-actualization. These theories expand the view of agency, highlighting the individual's autonomy and capacity for personal development beyond external stimuli or cognitive restructuring. Sociocultural theories (Vygotsky, 1978), in contrast, emphasize the role of social interactions and cultural context in shaping learning. They broaden the view of agency, suggesting that agency is socially mediated and influenced by collaborative interactions within cultural contexts.

With the numerous theories available to draw from, can a computational framework help to unify the theories and provide a pathway for new ideas in the field that advance the science of learning, including the integration of AI as a partner in research and learning? Perhaps a research framework that is friendly to computational representations needed for AI can also help to organize thinking about the scalability and homologous functions of intelligence as it traverses from cells to individuals, to groups, to world culture, despite the current shortcomings in understanding the complexity of human intelligence. As the AI researcher Fei-Fei Li has indicated, we do not need to know all the details of the complex mechanisms of AI to use its power to enhance our agency (Li, 2023) just as we do not need to know how drugs work to experience their benefits. In this chapter, we begin with the question of how to guide the development of AI-enhanced research into learning *processes*, for which a computational framework is a natural fit as the *lingua franca* of multidisciplinary researchers from machine learning, AI, and cognitive sciences who are interested in the future of AI in education.

1.3 Pros and Cons of a Computational Framework

A computational framework for learning theory could potentially provide the necessary foundation for modeling and simulating complex cognitive processes, allowing AI systems to mimic, understand, and augment human thinking and action. Its computational nature enables AI to process vast amounts of data, simulate cognitive mechanisms including multiagent experiments, and execute intricate learning algorithms, mirroring the way humans acquire knowledge and skills. Moreover, computational theories form the backbone for developing adaptive, learning-based AI systems capable of continuous improvement, adaptation to new information, and autonomous decision-making—essential qualities for effectively aiding human cognition and action across diverse tasks and contexts. Essentially, by grounding learning theories in a computational framework, AI developers can bridge the gap between theoretical understanding and practical application, ultimately advancing the potential for AI to enhance and augment human thinking and actions. We hope that this book provides food for thought to all who are interested in this topic.

Several objections could be raised against a computational theory of learning, challenging its applicability and explanatory power. Some computational approaches might oversimplify the complex and multifaceted nature of human learning, reducing it to algorithmic processes and neglecting the richness of human cognition and experience. To this objection, we emphasize that linear simplifications and mechanisms are indeed too simple, although in certain circumstances, reductionism, while incomplete, is still a useful tool of research. This book argues for a more complex nonlinear representational framework that includes insights from complexity and evolutionary theories. Second, some might argue that computational models might overlook the intricate biological underpinnings of learning, failing to capture the full complexity of neural processes and brain functions involved in learning and memory. The framework outlined in this book aims to show that with the addition of complexity and machine learning methods, research on learning can be biologically inspired and can add value by validating formerly non-computational ideas such as whether the addition of an expert in a social learning environment helps lead to expertise (Bhoopchand et al., 2023). A third objection to computational modeling of learning may be the prioritization of internal cognitive processes, neglecting the significant influence of social interactions, cultural contexts, and environmental factors that shape learning experiences. The tri-level model introduced and discussed here will directly answer this objection by integrating social and cultural learning processes interconnected to homologous functional roles in a causal model of individual learning processes.

Some may suggest that computational models might not adequately account for the role of emotions and creativity in learning, focusing primarily on information processing and neglecting affective and creative dimensions. To this objection, we point out that the creative expansion into the '*adjacent possible*' (Kauffman, 2000) provided by the causal learning cycle described in Chap. 6 is viewed as the foundation of all forms of creative physical, emotional, and cognitive growth. As will be explained in greater detail, Kauffman's point is that all evolutionary processes search, explore, and grow into the nearby space of possibilities. The conception of learning advanced here, involving emotional, creative, and cognitive mechanisms, is an evolutionary process at all three levels—individual, social, and cultural.

Some may also argue that computational models may struggle to capture implicit or unconscious learning processes that play a crucial role in knowledge acquisition and skill development. We point out that black boxes and unknown functions are straightforwardly included in our understanding of computational models, so the limitations of empirical concreteness and logical surety and completeness are not a barrier to the discovery and validation of new theory. These issues are part of the lessons of complexity that need to be included in the future of the learning sciences.

A final concern arises regarding the ethical implications of AI-based learning systems and the potential impact on human autonomy, privacy, and biases embedded in computational algorithms. We fully concur with the concern but do not place sole blame on computational methods for this issue. We maintain only that computation functions as an expressive language, facilitating precise communication and interaction with machines, and enabling complex tasks and problem-solving.

Considering the overlapping and conflicting ideas and objections, rethinking learning theory in the age of AI is required on several grounds: the partiality of description, which demands continued openness to new theories; the lessons of evolution and complexity, which demand continuous coadaptation of theory to reality; and more powerful computational models of learning, which deepen our understanding of prior theories and expand the reach and power of both symbolic and literal human activity.

1.3.1 The Partiality of Description

Descriptions within any theory or model face an inherent limitation or bias in how phenomena are understood or explained. No single theory or model can fully capture the complexity or entirety of a phenomenon, as the model must emphasize specific aspects or perspectives while neglecting others. This limitation inherently demands continued openness to new theories for several reasons:

- Existing theories might have blind spots or limitations in explaining certain aspects of phenomena. Embracing new theories allows for filling these gaps and expanding understanding. Such was the case when Einstein proposed that clocks and rods were not rigid measuring objects for space and time, which was a result of his special theory of relativity.
- New empirical data, technological advancements, or paradigm shifts might reveal aspects that challenge or extend existing theories, prompting the need for evolving knowledge through new frameworks. For example, the impact of 'large language models' on AI development has produced an enormous overflow of new ideas that are expanding the fields of computational psychology, machine learning, and agent-based experimentation. Also, synthetic data produced by simulations is increasing at the core of many sciences. The use of generative AI to create synthetic data is rapidly growing, relieving the burden of obtaining real-world data so machine learning models can be trained effectively. By the end of 2024, Gartner predicted 60% of data for AI will be synthetic to simulate reality, future scenarios, and de-risk AI, up from 1% in 2021 (icrunchdata, 2023).
- Different theories offer diverse perspectives or frameworks for understanding phenomena. Embracing new theories acknowledges the multiplicity of viewpoints, enriching understanding through increasingly diverse lenses. The book's 3-level model introduces a physical theory from a complexity theory of the biosphere into learning science and then discusses its implications.

- Phenomena are often multifaceted and dynamic. New theories might offer more nuanced or updated explanations that align better with the evolving nature of phenomena. For example, as more and more humans are interfacing with technology and personal agents, the frontiers of human-computer research are rapidly shifting from design processes led by humans into a coevolutionary stance, with machines occasionally taking the lead.
- New theories can provide context-specific or domain-specific insights that better suit particular applications or fields, enhancing their applicability. For example, in psychology, the emergence of coevolutionary theories has offered domain-specific insights into human behavior, enriching our understanding beyond traditional psychological frameworks.
- Finally, continuously being open to new theories prevents dogmatism or the stagnation of knowledge, fostering intellectual curiosity and growth. In what follows we introduce a new way to organize existing theories of learning that we hope will stimulate future research into cognitive systems, social systems, and cultural evolution.

In sum, the partiality inherent in descriptions and theories underscores the need for ongoing openness to new ideas. Embracing novel frameworks and perspectives enables a more comprehensive understanding, promotes intellectual evolution, and ensures that knowledge remains dynamic, adaptable, and better suited to grasp the richness and complexities of phenomena.

1.3.2 Tenets of Evolution and Complexity

Lessons derived from evolution and complexity theory also underscore the necessity for continual coadaptation of theoretical frameworks to an ever-evolving reality. Evolution, and its elaboration in complexity theory, highlights the natural role of learning as a part of elaborated exploration and filling of niches in a larger environment via incremental steps and leaps of progress. Evolutionary principles, for example, teach us that species survive and thrive by adapting to changing environments: organisms balance stability with adaptability in complex environments. So too, learning theories need to balance foundational stability while remaining open and adaptable to incorporate discoveries and shifts in understanding. Learning science theories must adapt to the evolving complexities and nuances of technological and social change to ensure continuing relevance and applicability.

In complex systems, interdependencies and emergent properties necessitate flexible and adaptable theories that accommodate new information and interactions among components. Just as organisms evolve through iterative processes of feedback and adaptation, theories benefit from constant refinement through feedback loops, incorporating new data and insights for improvement. Importantly, complexity theory embraces and explicates how nonlinear dynamics and unpredictability impact evolving systems. Learning theories must embrace the boundaries of uncertainty within the bounded infinity of complexity as well as the foundations of evolution and evolve to accommodate the coming unpredictable shifts driven by emergent phenomena.

Additionally, evolution and complexity often span multiple disciplines. For example, evolutionary biology's principles have been applied across disciplines to computer science and psychology, evident in evolutionary psychology's exploration of the adaptive origins of human behavior. Theoretical frameworks of learning science must also facilitate interdisciplinary synthesis; for example, including neuroscience, developmental psychology, social science, ethics, computer science, learning theory, and cultural transmission.

In summary, the tenets of evolution and complexity emphasize the need for learning theories to continuously adapt to the dynamic, interconnected, and unpredictable nature of research on learning. This continual evolution of theories helps ensure the relevance, applicability, and ability of learning science to provide meaningful insights into the complexities of learning processes in humans and machines.

1.4 Computational Models of Learning

Computational models of learning can enhance our comprehension of prior learning theories by providing mechanisms to simulate and test learning theories in practical controlled contexts, offering empirical validation and insights into their underlying processes. Computational models can handle both symbolic and literal signifiers of human learning by enabling the emulation of cognitive processes. Examples of this include a generative agent-based experiment at Stanford (Park et al., 2023) that created believable simulacra of human behavior, and an example from Google's Deep Mind (Bhoopchand et al., 2023) in which agents, using few shot imitation, succeed at real-time imitation of a human in novel contexts without using any pre-collected human data. The latter work identified a surprisingly simple set of ingredients sufficient for generating cultural transmission and developed an evaluation methodology for rigorously assessing it in a computational context. A third example is the development of a biology-based computing organoid interfaced with a silicon chip that illustrates how neurological network dynamics can outperform silicon for some network functions (Cai et al., 2023). Another example is a study that found large language models could replicate the behavior of individuals with underdeveloped cognitive and language skills (Milička et al., 2024). With human-like computational capabilities, computational models can foster the development of AI systems that mirror human-like learning, decision-making, and problem-solving, thereby amplifying the potential for advanced technological applications and augmenting human capabilities in a variety of domains.

Computational modeling is also needed to tackle combinatorial problems of too many variables and relationships. For example, pairwise testing, even of a small and completely connected network is not sufficient for building an understanding

(Bach & Schroeder, n.d) of a system. Even one or two continuous relationships among a set of variables can lead to uncertainty, complexity, and even chaos, which we next address with a hands-on example.

1.4.1 A Spreadsheet Model of Computational Modeling of Complexity

A simple spreadsheet model using Excel can help illustrate some of the core ideas of the computational modeling of complexity. The record-keeping and computational role in this example, while much less complex than in the AI discussed in the remainder of the book, is nonetheless helpful in cementing an intuition about dynamical systems—that a simple evolutionary rule can be the cause of many complex behaviors. This do-it-yourself exercise will provide a chance to introduce key terms that we'll need later and give you a chance to experiment with chaos and complexity to develop an intuition about the concepts that are central to the computational thesis and modeling framework. If playing with numbers and graphics doesn't interest you, this section can be skipped.

The model that follows is an example of a fully connected network with two nodes: an entity or agent in the 'now' and the same agent in the 'future.' The connecting link between the now and the future is a dynamic information signal that travels through a space of a bounded infinity or landscape of possibilities, illustrating what is meant by the *adjacent possible.* Starting with a new blank spreadsheet, type the following into the top left cell at A1 as an equation for a *dynamic* that will propel the *agent* across a *landscape* of numbers.

$$rx(1-x) \tag{1.1}$$

The equation has three terms or variables, *r, x, and (1−x).* The term *r* will be a real number between 1 and 4 (e.g. 3.789) representing the topology of the *behavioral landscape*, or you might think of it as the shape of the field of play or the affordance space for actions, for our *agent* to evolve. An agent is defined as any entity or process that has been empowered to act on behalf of another entity or process, in this case, the agent (x) acts based on its past self (1−x). We'll make later use of notions developed in this section (e.g. behavioral landscapes, agents, boundary conditions, bounded infinity, attractors) as we discuss resetting the terms in the framework for research in the learning sciences.

The variable *x* will be the agent's starting position in the behavior space at the beginning of the experiment, which will be a real number between 0 and 1. The interval from 0 to 1 is a *bounded infinity*, which means that there are an uncountable number of places the agent can go while remaining in the interval. The boundary conditions of 0 and 1 are limits to the small universe we are creating; numbers smaller or larger are undefined and not in the universe, yet there still are an infinite number of *x's* that our agent can become, which is what we mean by a bounded infinity. (Parallels to psychology, learning theory, and computer science will be discussed later). The term *(1−x)* is another bounded infinity, this time representing the *adjacent possible* position that our agent occupied in the last time step. Time is treated as a series of discreet steps so we can use the spreadsheet rows to experiment with the time course of the little system. Next, we'll finish the model, and then show a few experiments you can perform.

So far, we have stated in cell A1 the key *dynamic equation* propelling the state of the system over time, but we haven't made it operational, which we will do now. First let's label the columns A, B, and C with the three terms. In Cell A6 type *r*, cell B6 type *x*, and cell C6 type *(1−x)*. Just below these labels, we'll place the operational real numbers of the terms. The cells for your experimental inputs are A7, B7, and B8 (where the dynamic equation operates on the time step).

In cell B8 type this equation:

$$= \$A\$7^{*}B7^{*}\left(1-B7\right) \tag{1.2}$$

While Eq. 1.1 was a *pseudocode* of the dynamic, now sitting in cell B8 is the *computational equivalent* in a coded language that Excel understands. We will copy this code down the rows to produce *system behavior over time*. First, let's put in the starting values of our first experiment.

In cell A7 type the number *2,* in cell B7 type the number *0.9* (remember that we're staying in the real range from 0 to 1) and in cell C8 type = *B7* which copies the value of the previous time step into the current time. This emulates the idea that in the current time, there is a memory of the time just before, along with some change *caused* by the dynamic equation.

Now copy both B8 and C8 down to as many rows as you like. A couple of hundred rows will suffice for this example. You should have a sheet that looks somewhat like this (but we'll only show the first few rows) (Table 1.1).

Next, we can create graphic representations of the system to help develop our intuition of the system. We want to see how *x*, our 'agent' is changing over time, so we choose the y-axis as the range 0–1 and the x-axis as the 200 rows and get a result such as the left portion of Fig. 1.1.

Attractors come in three varieties—*point*, *cyclic,* and *chaotic*—and can motivate an agent to display several behaviors, which we will show in the next few examples. To illustrate the attractors, we create the *phase space portrait* of the system (see the right side of Fig. 1.1). In this graphic, we show the position of x at two positions: the current time and time step that just passed, by graphing columns B on the y-axis and C on the x-axis.

We are now ready to show different outcomes from the same dynamic equation (i.e. the rule that propels the state of our system to its next time step, which takes the current state *x* as a starting condition, factors in the shape of the landscape '*r*' and bounds the behavior to stay within a bounded infinity of affordances). The simplicity of the single rule is remarkable in that it can produce so much variety and

Table 1.1 A small model of chaos and complexity

	A	B	C
1	rx(1−x)		
2	r = fractal landscape dimension—real number between 1 and 4		
3	x = starting condition where 0 > x < 1		
4	(1−x) = the time step		
5			
6	**r**	**x**	(1−x)
7	2	0.9	
8		0.18	0.9
9		0.2952	0.18
10		0.41611392	0.2952
11		0.48592625	0.41611392
12		0.49960386	0.48592625
13		0.49999969	0.49960386
14		0.5	0.49999969
15		0.5	0.5
16		0.5	0.5
17		0.5	0.5
18		0.5	0.5
19		0.5	0.5
20		0.5	0.5
21		0.5	0.5
22		0.5	0.5
23		0.5	0.5

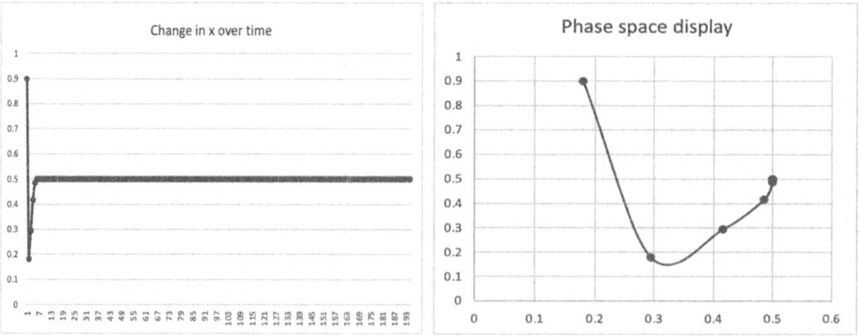

Fig. 1.1 Display of change in x over time and phase space display. Note: This system has a *point attractor* at .5 which the variable *x* reaches in step 15. In the phase space display, the attractor appears to act like a gravity well in the space, with the agent x evolving until it falls into the well

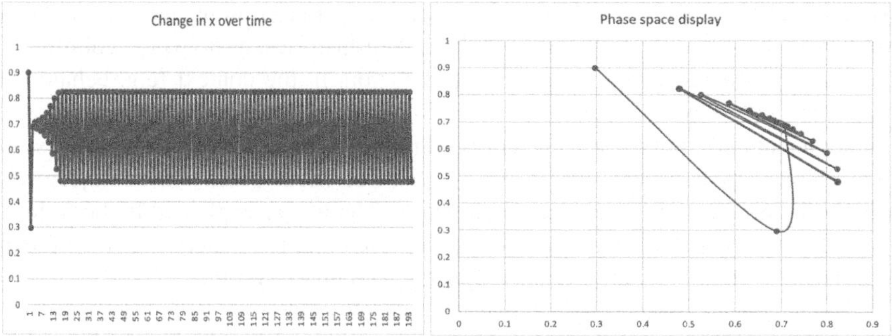

Fig. 1.2 Change in x over time and phase space display for r = 3.3. Note: With r = 3.3 the attractor is *cyclic*

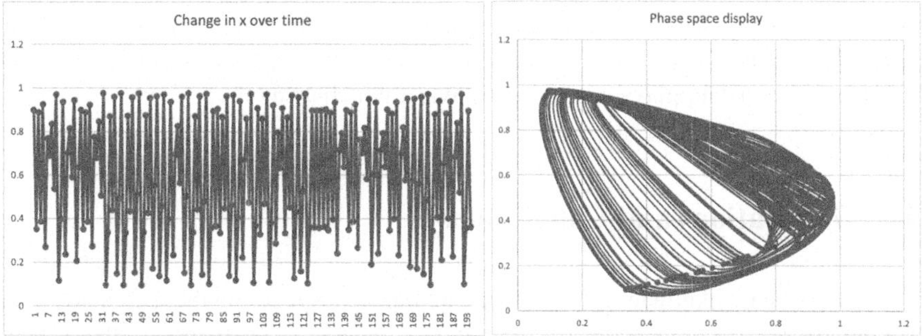

Fig. 1.3 Change in x over time and phase space display for r = 3.9. Note: With r = 3.9 the attractor is *chaotic*

uncertainty in the agent's behavior. For example, set *r* to 3.35 and see its phase space display in Fig. 1.2. To see an example of a chaotic attractor, set *r* to 3.9 and see its phase space displayed in Fig. 1.3. Note that in each of the examples thus far, the only variable we've been changing is the behavioral landscape (e.g. what we might think of as *context* in educational research). The underlying dynamic has not changed: the agent still factors in or *learns from* the same impact of time, and the contextual boundaries make sure that the new position remains in the same behavioral landscape. But remarkably, there are three very different regimes of behavior (point, cycle, and chaos) depending on the context.

This example has parallels to learning at all three levels (individuals, organizations, and cultures). The agent at the individual level is the learner, at the organizational level is the team, and at the cultural level is the whole evolving community with all its organizational and individual members. Individuals comprehend their current experiences by building on their past experiences, which can range from recognizing a familiar pattern (like the point and cyclic attractors) to needing to create new symbols and meaning from an experience that is more chaotic and uncertain (like the chaotic attractors). Similarly, teams and organizations also comprehend and create based on (and often bounded by) their histories, as do whole cultures.

We can now see that a simple rule that factors in the context (e.g. behavioral landscape complexity), starting point (e.g. a previous position in the state space), and boundary conditions (e.g. the bounded infinity of a universe of affordances) of a dynamical system can cause a wide variety of agent behaviors or trajectories shaped by point, cyclic, and chaotic attractors. The theory explained in the following chapters treats learning as a result of a more complex underlying rule system that is contextually bounded yet also dynamic and highly varied.

You may wish to experiment with other inputs to explore the impacts of the starting position and the complexity of the behavioral landscape. Try $r = 3.85$ to see a cyclic attractor with clear waypoints on the cycle. Another avenue of exploration is the complete behavior space of this small model, which is known as the 'route to chaos'- the process by which a simple attracting set for a dynamical system becomes chaotic as an external parameter is varied. The Feigenbaum constant $\delta \simeq 4.669$ measures the convergence rate of different iterated maps and is a universal feature of the approach to chaos in a class of systems. There is much more to discover concerning the role that constant plays in learning science, which awaits!

1.5 Limitations and Conclusions

This chapter has acknowledged that there are numerous learning theories in the field but did not attempt to provide a comprehensive analysis or evaluation of these theories. Instead, specific limitations or blind spots of existing learning theories were discussed as explaining only part of the complex phenomenon, as each theory focuses on specific facets of the learning process, resulting in overlapping ideas. Active agency and participation, for example, are common concerns in most theories. However, future learning theories must be *computational* theories for artificial intelligence to enhance human learning processes, thinking, and action. This implies that AI should use transparent and understandable human communication to fulfill its promise as an enhancement to human learning. Embracing new theories and diverse perspectives will enrich understanding and facilitate more precise communication and interaction with AI agents. While some computational models of learning may overlook for example, the complexity of human cognition and experience, the biological underpinnings of learning, the influence of social interactions and cultural contexts, or the role of emotions and creativity, the addition of complexity and machine learning methods to learning theory may be able to address these limitations and provide biologically inspired research on learning.

The chapter has emphasized the need for learning theories to continuously co-adapt to the dynamic, interconnected, and unpredictable nature of research on learning. It asserted that the continual evolution of theories is important to ensure the relevance, applicability, and ability of learning science to provide meaningful insights into the complexities of learning processes involving both humans and machines. Future research, the chapter established, should focus on organizing existing theories of learning in a new way, to stimulate research into a model that integrates cognitive systems, social systems, and cultural evolution.

1.6 Moving Forward

The role of the computer in the spreadsheet model is a simple computing and record-keeping role (e.g. multiplying factors and displaying results), it was also helpful in cementing an intuition about dynamical systems—that a simple evolutionary rule can be the cause of many complex behaviors. Chapters ahead are going to introduce a causal dynamic into learning theory and attempt to integrate new ideas with existing learning theories by situating them in a foundation that spans from individual learning processes to group learning processes, to cultural learning processes. Because the causal dynamic is fundamentally a computation, and that is the language of AI, the move necessitates a computational learning theory framework to advance and leverage AI to promote learning processes at every level. At this point, the simple role of the computer used in this chapter is metaphorically only the first expression of a longer more elaborated narrative about the role of computation in AI, general intelligence, and learning processes, which we believe underscores its need for inclusion in the future of the learning sciences.

Hopefully, we have opened the conversation sufficiently to indicate the saliency of questions about the capabilities of computing power, when coordinated with theories of complexity, dynamical systems, and evolution, is important to developing the potential roles of AI in the learning sciences. Those questions include generative and creative roles in learning, teaching, research, and education systems. In the chapters that follow, we will focus on assisting learning processes at a variety of levels utilizing a new synthesis of learning theories integrated with computational methods involved in AI. This leads us to reiterate that learning theories from this point forward must be computational theories if artificial intelligence is to fulfill its promise as an enhancement to human learning processes, thinking, and action.

References

Bach, J., & Schroeder, P. J. (n.d.). *Pairwise testing: A best practice that isn't*. https://bbst.courses/wpcontent/uploads/2022/08/Bach_PairwisePNSQC2004.pdf

Bhoopchand, A., Brownfield, B., Collister, A., Dal Lago, A., Edwards, A., Everett, R., Fréchette, A., Oliveira, Y. G., Hughes, E., Mathewson, K. W., Mendolicchio, P., Pawar, J., Pîslar, M., Platonov, A., Senter, E., Singh, S., Zacherl, A., & Zhang, L. M. (2023). Learning few-shot imitation as cultural transmission. *Nature Communications, 14*(1), 7536. https://doi.org/10.1038/s41467-023-42875-2

Brown, J., Collins, A., & Duguid, P. (1989). Situated cognition and the culture of learning. *Educational Researcher, 18*(1), 32–42. https://doi.org/10.3102/0013189X018001032

Cai, H., Ao, Z., Tian, C., Wu, Z., Liu, H., Tchieu, J., Gu, M., Mackie, K., & Guo, F. (2023). Brain organoid reservoir computing for artificial intelligence. *Nature Electronics*. https://doi.org/10.1038/s41928-023-01069-w

icrunchdata. (2023, December 14). 5 trends shaping the future of data science and machine learning. *Icrunchdata*. https://icrunchdata.com/blog/5-trends-shaping-the-future-of-data-science-and-machine-learning/

Kauffman, S. (2000). *Investigations*. Oxford University Press.

Kolb, D., & Fry, R. (1975). Towards an applied theory of experiential learning. In Cooper (Ed.), *Theories of group processes* (pp. 33–38). Wiley.

Lave, J. (1991). Situating learning in communities of practice. In L. B. Resnick, J. M. Levine, & S. D. Teasley (Eds.), *Perspectives on socially shared cognition* (pp. 63–82). American Psychological Association. http://scholar.google.com.hk/scholar?q=communities+of+practice&hl=zh-TW&lr=#7

Li, F. (2023). *The worlds I see: Curiosity, exploration, and discovery at the dawn of AI. Moment of Lift Books*. Flatiron Books.

Maslow, A. H. (1943). A theory of human motivation. *Psychological Review, 50*(4), 370–396. https://doi.org/10.1037/h0054346

Milička, J., Marklová, A., VanSlambrouck, K., Pospíšilová, E., Šimsová, J., Harvan, S., & Drobil, O. (2024). Large language models are able to downplay their cognitive abilities to fit the persona they simulate. *PLoS One, 19*(3), e0298522. https://doi.org/10.1371/journal.pone.0298522

Park, J. S., O'Brien, J. C., Cai, C. J., Morris, M. R., Liang, P., & Bernstein, M. S. (2023). *Generative agents: Interactive simulacra of human behavior*. http://arxiv.org/abs/2304.03442

Rogers, C. R., Dorfman, E., Gordon, T., & Hobbs, N. (2015). *Client centered therapy: Its current practice, implications and theory* (Reprinted). Robinson.

Sweller, J., Ayres, P., & Kalyuga, S. (2011). *Cognitive load theory*. Springer. https://doi.org/10.1007/978-1-4419-8126-4

Vygotsky, L. S. (1978). *Mind in society: The development of higher psychological processes*. In M. Cole, S. John-Steiner, S. Scritmer, & E. Souberman (Eds.), Harvard University Press.

Zhou, M., & Brown, D. (2015). *Educational learning theories* (2nd ed.). GALILEO, University System of Georgia.

Chapter 2
AI and Learning Processes

2.1 Introduction

As many modern fields of inquiry and expression are increasingly underpinned and enhanced by computational resources, there is a common ground for a unifying framework encompassing artificial intelligence, learning analytics, educational data mining, machine learning, and complexity theory (Dawson et al., 2018; Tsai et al., 2019). In Chap. 1, we explored several of the intersections of these fields and introduced key terms for a discussion of complexity and causation in a computational theory. The aim was to build a case for the inclusion of computation as a requirement for learning theories intended to advance the potential of AI in education. In this chapter, we explain why developing the potential of AI should start from *learning processes* rather than the many other facets of education where AI is likely to play a role in the coming years.

Educational systems are wide-ranging, diverse collections of solutions to the challenge of transmitting cultural knowledge, fostering citizenship, cultivating critical thinking skills, and preparing individuals for active participation in society. Recently, remarkable new AI capabilities have become available, with dramatic potential impact on society. For example, a deep learning AI algorithm can now detect childhood autism with 100% accuracy from an image of the pupil of the eye (Kim et al., 2023) and deep learning-guided discovery of structural classes of antibiotics has demonstrated that machine learning models in drug discovery can be explainable, providing insights into the chemical substructures that underlie selective antibiotic activity (Wong et al., 2023). In addition, the first discoveries made for established open mathematical problems using large language models (LLMs) were recently published in (Romera-Paredes et al., 2023).

In scholarship and publications, AI systems are now increasingly used to help write articles and essays, outline papers, produce original pieces of art, and act as collaborators in academic research. In state-of-the-art machine learning approaches

© The Author(s), under exclusive license to Springer Nature Switzerland AG 2024

D. C. Gibson, D. Ifenthaler, *Computational Learning Theories*, Advances in Analytics for Learning and Teaching,
https://doi.org/10.1007/978-3-031-65898-3_2

such as LLMs (Huang et al., 2022; Zhou et al., 2022) an AI agent may be responsible for much of the intellectual work in these activities, which raises several questions about the role of AI in education. Should a higher education institution grant a degree to a person using such a system? Should an employer hire someone based on what they know and know how to do—with or without AI? Would either of these cases be an example of a potential unfair practice, or 'cheating?' What should happen if a person gets identified for cheating by an algorithm and an unknown professor (Hill, 2022)? Who will be responsible if a person is trapped into making important decisions by intelligent systems that use hidden rules for identifying and addressing people in poverty (Hao, 2022)?

These areas surrounding the new capabilities and attendant concerns of AI in education all need dedicated basic research, design research, creation of interventions, and field study, but they are endpoints and effects rather than a locus for the development of AI to advance the science of learning, which we believe should start with the processes of individual, organizational and cultural learning.

2.2 Learning Processes as the Key Leverage Point

While aspects of education, such as schooling, administration, curriculum, teaching, and technology, are essential components of the educational landscape, learning *processes* should take precedence in learning science due to their foundational role in shaping educational experiences and outcomes at every level. If the new framework outlined here proves useful for individual, team, and organizational learning, then the other components of education are likely to evolve effective supportive structures and practices.

More central to learning science than the details of educational delivery, the core physical process of learning is *a latent causal driver of the dynamics of the system at every level*. By latent driver, we mean that the relevant variables are often hidden from direct observation, or they are constructs hypothesized to facilitate scientific explanation. The equation in the spreadsheet model in Chap. 1 ($rx(1-x)$) was such a latent driver of the dynamics of that system. In the 3-tier model introduced in this book, the focus is on the future discovery of dynamic equations that exist at every level and represent key learning processes. This idea is advanced for several reasons.

First, learning processes encapsulate *how learning occurs*. The processes capture and utilize the key mechanisms, cognitive processes, and neuropsychological principles underlying how individuals acquire, retain, and apply knowledge. Key questions here for exposition in the remaining chapters are to what extent and how those processes scale and adapt beyond individuals to teams and larger organizations and society.

Second, learning processes also have a direct impact on outcomes, because they profoundly influence how individuals comprehend, retain, and utilize information, directly impacting academic achievement, joining teams and organizations, and lifelong learning. Later chapters will explore how these direct impacts of individual learning scale and adapt to teams, organizations, and society.

Third, learning processes are the basis for employing evidence-based pedagogical strategies, optimizing instructional methods, and facilitating more effective teaching practices. For example, insights into learning processes are required for tailoring and personalizing learning experiences to accommodate the diverse needs of individual learners. Can similar evidence-based roles be found for the learning processes of collaborating teams and for tailoring pedagogical interventions for organizations or larger groups within society? If so, a scalable, adaptable framework of learning processes is needed for the design and implementation of AI-enhanced educational technologies that are well-aligned with cognitive and psychological principles for increased efficacy.

2.3 Introducing the New Framework

The new framework introduced here and detailed further in the following chapters is a multidisciplinary integration and expansion of existing learning theories, which are unified by literal and functional similarities in models from intersecting fields of research implicated in learning, including complex systems, biology, psychology, computer science, sociology, and pedagogy (Fig. 2.1). The model brings together three historical sources: theories of learning enhanced by complexity considerations to form a core causal learning mechanism (Micro Level—Chaps. 3, 4, 5, and 6), a cognitive and social science model of how people learn (Meso Level—Chaps. 7, 8, 9, 10, and 11), and cultural-historical activity theory (Macro Level—Chaps. 12 and 13).

The 3-level framework is offered in hopes of leading to a better understanding of the micro, meso, and macro levels of AI-assisted learning and the implications for developing new ways of generating, collecting, analyzing, and interpreting digital learning experience impacts when AI is an active coagent of learning. At the micro level, AI is an increasingly intelligent partner in improving a learner's emotional, behavioral, and cognitive capabilities; and these new agential possibilities of AI acting on key traditional individual learning theory constructs are implied by and entailed by the core learning process. At the meso level, cross-disciplinary ideas in the framework expand upon existing theories of the nature and roles of knowledge, learner, community, and feedback to become interacting mechanisms, states, and processes in group learning enhanced by AI. At the macro level, multidisciplinary and cross-cultural learning mechanisms are integrated with new subnetwork models of community and individuals.

The three tiers—individual, organizational, and cultural—need to be integrated because of the overlapping membership relationships of learners to teams and organizations in society at large, and the role of context in understanding learning processes. Contexts cover a wide range; from the potential of a single individual to make a world-changing societal contribution to the limitation on learning of being

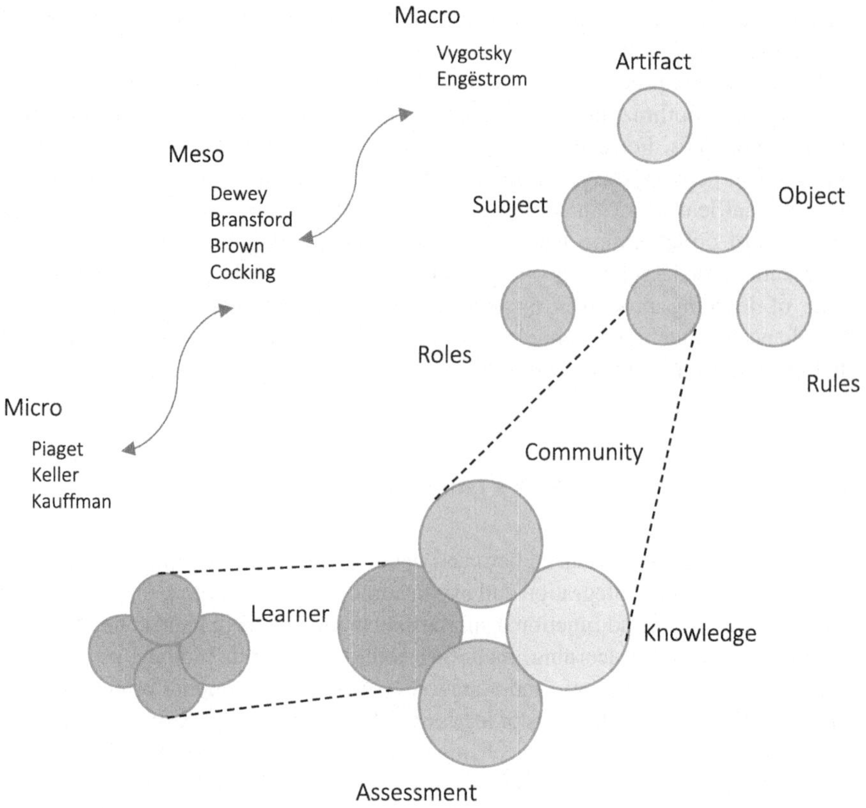

Fig. 2.1 3-tier model of learning processes. Note: The 14 nodes of the 3-tier network model are supported by a variety of theories of individual, organizational, and cultural learning. Linking relationships among the nodes are defined and discussed as causal influences dynamically advancing states of the network in part or in whole, with context defined as all remaining nodes of the network in an analysis

from a particular family or culture or possessing a particular set of psychological predispositions. The challenge of integrating existing theories into the new computational framework is both wide in terms of the heterogeneity of encompassing concepts and deep in terms of the scale of the hidden layers of interpenetrating micro-to-macro processes. Fortunately, AI has evolved to the point of being a highly capable partner in researching and developing an understanding of complex multi-tiered systems.

Understandably, with the new possibilities for multidisciplinary research on AI in education rapidly evolving, the proposed framework is in many respects tentative and expects to be expanded on by others as it is inclusive of existing theories, bridges gaps, and explores new ideas. The framework is offered as a discussion launchpad rather than a definitive and finalized answer to all identified challenges. Nevertheless, while acknowledging the need for a critical review of the proposed framework, the chapters that follow offer a new synthesis of ideas for thinking about AI as:

- A partner in exploration, learning, and expression by an individual
- A creative, collaborative group member in team activities
- A mediator as well as an initiator of cultural shifts

These roles are emerging because AI in education is changing what it means to know and do things in concert with intelligent computational resources.

2.4 Potential Pitfalls of AI Interventions in Education

In this concluding section, we briefly review and critique AI's impacts on other entry points to education beyond the core processes of learning. From a technical perspective, AI-driven adaptive learning platforms seem poised to increasingly personalize education by tailoring content and pacing to individual student needs, enhancing learning outcomes. This shift may disrupt current systems of licensing, schooling, and validation of learning. AI is also poised to improve the analysis of educational data to inform policies, curriculum design, and resource allocation, optimizing educational strategies based on empirical insights. AI will also reduce workloads and bring new efficiencies by automating administrative duties like grading, scheduling, and resource management, streamlining operations, and reducing workloads. However, these entry points might overlook the human side of the challenges, the political realities of the change process, and the symbolic shifts caused by the increasing presence of AI in society.

Among the key human resource issues are teacher capabilities including technological literacy. AI is likely to cause shifts in pedagogical paradigms as it alters teaching approaches and empowers student-centric learning, which could transform the perception of effective teaching. Increased reliance on AI technologies by both teachers and students may reshape perceptions of technological literacy and critical faculties, for example, skills needed to recognize deep fakes in media and to move away from relying on one-shot assessment products and processes that an AI can pass better than humans (Schubert et al., 2023). The profession of teaching may change as AI can provide aids and personalized teaching assistance to anyone at any time; so, what will be the value-add of a teacher? AI-powered training and resources may indeed alleviate some parts of traditional teacher education. Educator roles could shift to focus more on mentoring, individualized instruction, and leveraging AI tools to support diverse student needs, as AI-enhanced instruction increasingly can teach anyone about anything at any time. These changes in the delivery systems of education may look entirely different after AI is ubiquitously integrated, but these too seem like outcomes rather than the drivers of learning science research applied to developing the requisite AI.

AI will undoubtedly also exert political influences on educational policymaking, requiring regulations and guidelines on AI integration, data privacy, ethics, and access to AI-driven resources. AI's impact on education may exacerbate educational disparities, necessitating policies to ensure equitable access to AI-driven tools and resources for all students. To discuss these cases would venture into prescriptive advice, guidelines, or rules, which, while valuable, are beyond our intended scope focusing on basic science. The reader is directed to others (Dieterle et al., 2022) for a discussion of the ethical effects of AI in education and for advice on creating a virtuous cycle of effects on access, representation, algorithms, interpretation, and citizenship that need to be watched as AI is used in education.

In addition to potential harms, educators working to develop AI solutions may also be vulnerable to the over-hyped potentials and pitfalls of AI in education—'AI theater' according to (Selwyn, 2022) perhaps fed in part by 'enchanted determinism' (Campolo & Crawford, 2020)—the sense that technology will somehow rise magically with superhuman powers to address education's shortcomings or save humanity from its worst impulses. Our narrative will attempt to avoid these extremes of wishful thinking or alarmist rhetoric by acknowledging that AI and its global big data foundation are enhancements and distributions of human thinking and performance that magnify human potential.

However, as a magnifier, AI can make the bad worse or the good better and must be developed and used in education with that understanding in mind. All tools and artifacts with the power of enhancement of human potential mediate the intentions of the people employing them (Engeström et al., 1998; Vygotsky, 1978). For example, steam engines, and later oil and gas burning engines, reduced labor costs for moving vast quantities of earth and made it possible to build enormous dams and cities. However, they also led to displacements, injuries, and unemployment. AI in education is undoubtedly a double-edged sword that will lead to unintended consequences and likely cause a rethinking of many assumptions about learning, knowledge, skill, performance, creativity, and innovation. Critical to its positive use is the intention of whoever wields power, with appropriate caution and watchfulness for consequential impacts.

2.5 Limitations and Conclusions

This chapter hinted at some of the potential harms and pitfalls of AI in education, which are important to consider when developing AI solutions, but did not extensively discuss the issues. The chapter broadly discussed the need for a unifying framework that encompasses artificial intelligence, learning analytics, educational data mining, machine learning, and complexity theory, highlighting the interconnectedness of these fields. The chapter emphasized the importance of starting the development of AI in education from *learning processes*, as they play a foundational role in shaping educational experiences and outcomes, and it asserted that the core processes of learning drive the dynamics of the system at every level (e.g. micro, meso, and macro). The chapter emphasized the potential importance of

AI-enhanced learning processes driving the evolution of effective educational practices.

When AI is an active coagent of learning it entails and supports new ways of generating, collecting, analyzing, and interpreting data for the science of learning. While future studies may also consider additional aspects such as the human side of AI in education, the political realities of the change process initiated by the new AI capabilities and roles in education, and the symbolic shifts caused by the increasing presence of AI in society, ultimately the theories will need to connect to *learning processes* embedded at every level to explain how people learn with AI.

2.6 Moving Forward

In Chap. 3 the nodes and relationships of a 3-level network theory of micro-meso-macro learning processes are introduced.

References

Campolo, A., & Crawford, K. (2020). Enchanted determinism: Power without responsibility in artificial intelligence. *Engaging Science, Technology, and Society, 6*, 1–19. https://doi.org/10.17351/ests2020.277

Dawson, S., Poquet, O., Colvin, C., Rogers, T., Pardo, A., & Gasevic, D. (2018). *Rethinking learning analytics adoption through complexity leadership theory* (pp. 236–244). https://doi.org/10.1145/3170358.3170375

Dieterle, E., Dede, C., & Walker, M. (2022). The cyclical ethical effects of using artificial intelligence in education. *AI & Society*. https://doi.org/10.1007/S00146-022-01497-W

Engeström, Y., Miettinen, R., & Punamäki, R. (1998). *Perspectives on activity theory*. Cambridge University Press.

Hao, K. (2022). *The coming war on the hidden algorithms that trap people in poverty*. MIT Technology Rview. https://www.technologyreview.com/2020/12/04/1013068/algorithms-create-a-poverty-trap-lawyers-fight-back/

Hill, K. (2022). Accused of cheating by an algorithm, and a professor she had never met. *New York Times*. https://www.nytimes.com/2022/05/27/technology/college-students-cheating-software-honorlock.html

Huang, J., Gu, S. S., Hou, L., Wu, Y., Wang, X., Yu, H., & Han, J. (2022). *Large language models can self-improve*. http://arxiv.org/abs/2210.11610

Kim, J. H., Hong, J., Choi, H., Kang, H. G., Yoon, S., Hwang, J. Y., Park, Y. R., & Cheon, K.-A. (2023). Development of deep ensembles to screen for autism and symptom severity using retinal photographs. *JAMA Network Open, 6*(12), e2347692. https://doi.org/10.1001/jamanetworkopen.2023.47692

Romera-Paredes, B., Barekatain, M., Novikov, A., Balog, M., Kumar, M. P., Dupont, E., Ruiz, F. J. R., Ellenberg, J. S., Wang, P., Fawzi, O., Kohli, P., & Fawzi, A. (2023). Mathematical discoveries from program search with large language models. *Nature*. https://doi.org/10.1038/s41586-023-06924-6

Schubert, M. C., Wick, W., & Venkataramani, V. (2023). Performance of large language models on a neurology board–style examination. *JAMA Network Open, 6*(12), e2346721. https://doi.org/10.1001/jamanetworkopen.2023.46721

Selwyn, N. (2022). The future of AI and education: Some cautionary notes. *European Journal of Education*. https://doi.org/10.1111/EJED.12532

Tsai, Y.-S., Poquet, S., Gašević, D., Dawson, S., & Pardo, A. (2019). Complexity leadership in learning analytics: Drivers, challenges, and opportunities. *British Journal of Educational Technology, 0*(0), 1–16. https://doi.org/10.1111/bjet.12846

Vygotsky, L. S. (1978). *Mind in society: The development of higher psychological processes*. In M. Cole, S. John-Steiner, S. Scritmer, & E. Souberman (Eds.), Harvard University Press.

Wong, F., Zheng, E. J., Valeri, J. A., Donghia, N. M., Anahtar, M. N., Omori, S., Li, A., Cubillos-Ruiz, A., Krishnan, A., Jin, W., Manson, A. L., Friedrichs, J., Helbig, R., Hajian, B., Fiejtek, D. K., Wagner, F. F., Soutter, H. H., Earl, A. M., Stokes, J. M., et al. (2023). Discovery of a structural class of antibiotics with explainable deep learning. *Nature*. https://doi.org/10.1038/s41586-023-06887-8

Zhou, Y., Muresanu, A. I., Han, Z., Paster, K., Pitis, S., Chan, H., & Ba, J. (2022). Large language models are human-level prompt engineers. *arXiv*. http://arxiv.org/abs/2211.01910

Chapter 3
A Complex Hierarchical Framework of Learning

3.1 Introduction

Frameworks for learning theory have been offered from several disciplines including psychology (e.g. behaviorism, cognitivism, and constructivism), biology (e.g. autocatalysis, evolution, emergence), and computer science (e.g. one-shot learning, deep learning, neural networks, infinite mixed membership stochastic block models) but these theories have rarely been integrated or unified. Treating the frameworks as unrelated to one another is adequate for modeling and conducting inquiries that further refine each perspective, but do not help advance cross-disciplinary knowledge or understanding and do not contribute to understanding the complexities arising in the intersections of the domains.

For instance, consider the potential intersection between neural networks in computer science and theories of emergence in biology. One could explore how the emergent properties observed in biological systems might inform the development of more adaptive and resilient neural network architectures. Or vice versa, one could explore how principles from artificial neural networks might provide insights into emergent behaviors such as learning new skills. Integrating various frameworks to promote learning processes could unlock profound insights into the complex interplay between biological, psychological, and computational aspects of learning, offering a richer understanding of learning mechanisms and fostering innovations at the intersections of these domains. In this chapter, we provide a framework for this kind of exploration and in later chapters, we will present concrete ideas about artificial intelligence (AI) that focus on the integration of theories to promote learning processes at all levels.

Consider, for example, the potential intersection between autocatalysis—a fundamental concept in biological systems—and theories of learning. Autocatalysis is a physical process that leads to self-sustaining chemical reactions and the emergence of complex systems. In the realm of learning, this concept offers insights into

© The Author(s), under exclusive license to Springer Nature Switzerland AG 2024

D. C. Gibson, D. Ifenthaler, *Computational Learning Theories*, Advances in Analytics for Learning and Teaching,
https://doi.org/10.1007/978-3-031-65898-3_3

self-reinforcing learning processes, particularly since we are introducing a model of the physical mechanisms of learning integrated with psychological, social, and cultural facets. The complex hierarchical theory implies that learning mechanisms within individuals and groups exhibit autocatalytic properties, where initial knowledge or skills catalyze the acquisition of further knowledge or skills in self-reinforcing loops aligned with the causal loop model outlined in Chaps. 4, 5, and 6 and then applied to AI application development for individual learning in Chap. 7, social learning in Chaps. 10 and 11, and cultural learning in Chap. 13. Just as some molecules catalyze the environment and processes of their self-production, initial learnings catalyze the acquisition of related and more complex knowledge, leading to a self-sustaining cycle of learning and knowledge growth. Understanding the parallels between the fundamentals of autocatalysis in biological systems and the self-reinforcing mechanisms in learning might reshape how we perceive and design educational strategies, emphasizing the importance of initial learning experiences and their impact on subsequent knowledge acquisition and skill development. Exploring this intersection in subsequent chapters will hopefully pave the way for novel approaches to fostering efficient and continuous learning processes enhanced by artificial intelligence (AI).

AI has broken through to educational research as a powerful generative source of novel information since the widespread public awareness and use of 'large language models' (LLMs) appeared in late 2022 (Dai et al., 2023; Huang et al., 2022; Wei et al., 2022; Zhou et al., 2022). Two examples of the breakthrough shifts are AI in adaptive learning systems and automated assessment and feedback. AI-powered adaptive learning systems have become more sophisticated, leveraging large language models to offer personalized tutoring experiences that embody a better understanding of students' queries and learning patterns, allowing for more nuanced and tailored educational content delivery. AI models can now automate grading processes for written assignments and provide detailed feedback that more comprehensively analyzes students' responses, offering not just scores but also specific feedback akin to that of a human instructor.

LLMs have also significantly facilitated cross-disciplinary research by providing advanced language understanding and data analysis capabilities. For example, in biomedical research LLMs are used to process and analyze vast amounts of medical literature, extracting valuable insights, identifying patterns, and accelerating the discovery of new treatments or associations within the field. In healthcare, AI now aids in understanding patient records, enabling better clinical decision support, patient outcome predictions, and the extraction of relevant information from diverse medical sources (Giordano et al., 2021). In climate science, LLMs help interpret environmental texts, policies, and reports across multiple languages, facilitating cross-border collaborations and a comprehensive understanding of global environmental issues (Kumari & Pandey, 2023).

These examples highlight how large language models have become instrumental in breaking down barriers between disciplines. By offering advanced natural language understanding and processing capabilities, AI can now assist researchers in gaining insights, interpreting vast amounts of information, and fostering collaborations across diverse fields such as biomedicine, climate science, and beyond. However, there is a lack of unifying and human-centered ideas for guiding the development of AI as well as for dealing with the many new challenges arising precisely due to the methods of training, shaping, and utilizing its power to promote learning and prioritize human concerns. So, a word of caution is in order. AI today tends to reproduce and magnify or amplify biases present in the data it has been trained on, and AI decisions are often opaque and hard to interpret or explain, which can be especially troublesome when errors occur on critical matters (Jones, 2023). Addressing these kinds of challenges necessitates the development of unified principles and guidelines that prioritize human concerns, ensuring that AI's power to promote learning aligns with ethical considerations and transparency, fostering trust and accountability in its utilization.

If current trends continue, artificial intelligence will more fully participate and be embedded in the distributed intelligence of individuals and groups. A sign of this imminent future can be found in advice recently given by publishers concerning the declaration of the use of AI in research and writing. Not only are authors now required to acknowledge the use of AI, but so are reviewers! This illustrates the increased ability of AI to read, interpret, and make judgments, with the result that publishers want to be sure reviewers know that the ultimate responsibility for a review remains with the human (Nature Editors, 2023). Additionally, as AI approaches higher and higher levels of 'general intelligence,' learning at all levels is likely to be impacted, as we will outline in this book in more than a dozen specific ways. We hold that the increasing intelligence of AI will require a learning sciences framework that supports research into, and understanding of, a complex hierarchy of learning systems, with terms we will define in this chapter.

In Chap. 1 we explained why learning theories must be *computational* learning theories from this point forward if artificial intelligence is to fulfill its promise as an enhancement to human thinking and action. In Chap. 2 we focused on promoting learning *processes* rather than on using AI to develop or shape content, assist with teaching, support educational leadership and policy decision-making, or reach other societal aims of education, to center attention on AI assisting the core mechanisms of human learning by individuals, groups and culture at large. In this chapter, the goal is to introduce a proposed three-tiered model of learning processes and define its foundational ideas, which will then be used as the foundation for the rest of the book.

In the remainder of the book, we provide details of the system framework of micro—meso—macro levels, which we view as a comprehensive system of learning that spans from individuals to teams and organizations, and to society. At each level of the system are agents (e.g. semiautonomous learning processes and mechanisms) and entities (e.g. individual learners, teams, and organizations) connected to and influencing each other, which necessitates that we focus on *network modeling* of the *entities*, *agents,* and *dynamics*. We conceptualize the 3-tiered model as a multiply-connected system represented by a *network of networks* (Fig. 3.1).

Each tier comprises a distinct set of entities and agents that interact within their respective domains, yet their influence also transcends the tier boundaries, creating

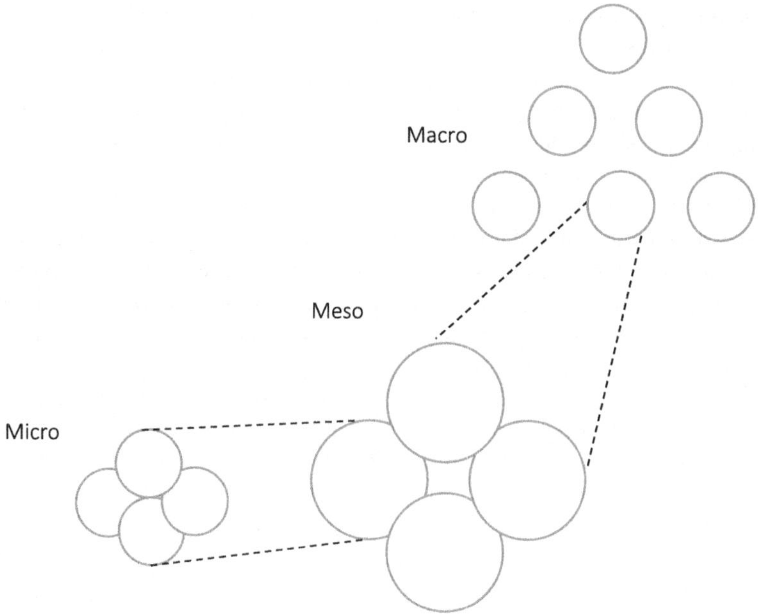

Fig. 3.1 A complex hierarchical system model for a computational learning theory. Note: The micro-level (individual learner) mechanisms become approximate (analogous and homologous) replications multiplied at the meso level (learners join teams, organizations, and community activities) in an expanded context of team-based knowledge building. At the macro level, multiple teams become approximately replicated and multiplied into larger cultural entities such as schools of thought and fields of practice comprised of interdisciplinary and international community cultures

a web of interconnectedness across the entire system. The multicultural milieu of the macro network of 'the world at large' in our view encompasses all meso-level networks of human organizations and smaller teams of which individuals (with internal micro networks) are members. It is perhaps obvious that this characterization requires a *dynamical hierarchical model* since networks of cultures contain networks of organizations and teams, which in turn contain individuals, themselves networks of distributed and embodied intelligence, and all entities are evolving and learning over time.

A core learning process, which we introduce here as the *causal learning cycle,* evolves and is maintained at each level, with appropriate homology to each level of the system. By 'causal' we mean to imply that a process or outcome necessarily (or with overwhelming probability) entails or leads to the next phase in the model, as detailed in Chaps. 4, 5, and 6. By 'homology' we mean to say that similar relations, relative positions, or structures exist at the micro, meso, and macro levels that closely replicate and expand upon the dynamics of the enclosed subsystems. For example, similar processes of learning occur in individuals and teams, as described in Chap. 9.

Of note, at every level of the micro-meso-macro system and key to the learning cycles within each level, are self-reflecting feedback and feed-forward connections. These reentrant cycles of information give rise to *complexity* in the system (by bounding the infinite), which we will endeavor to explain and illustrate. The defined entities (e.g. the nodes representing people and processes) and relationships of the network (e.g. the links holding the dynamic updating rules of the system) are then used to define specific roles for AI to play in the learning processes at each level. We will now define and elaborate on the key foundational ideas and briefly outline the framework, which will be elaborated on in the remaining chapters.

3.2 Learning Processes at the Micro, Meso, and Macro Levels

First, we introduce some examples of the entities and processes at each level. We will later fully define these entities in detail, not only as they initially arose in historical theories of learning and change, but also which are now connected to a new dynamic structure that includes a causal chain of events, which we assert is present in a variety of forms in each level.

Level 1: Micro-level *individual learning processes* include paying attention to something arising out of the equilibrium of the status quo, figuring out if it is something new or not, fitting in the new awareness to an experience or an existing schema, or making an accommodation to create or change a mental model, and satisfying oneself that something has been understood or recognized.

Level 2: Meso-level *group learning processes* include learning from others, forming or becoming associated with a team, sharing problems and concerns with others, reaching an understanding with others, receiving feedback from a person or group, and bringing one's unique perspective into contact with other people's points of view.

Level 3: Macro-level *interdisciplinary cultural learning processes* include transferring knowledge and techniques from one field to another, redefining problems using terms and ideas from more than one field, changing or evolving the knowledge base of a field of endeavor, and admitting new communities of practice into a larger setting that combines and acknowledges alternative ways of knowing, being and becoming.

Individuals within groups and learning community groups within cultures are examples of *embedded motifs*, which in network theory, refer to recurring and specific patterns or substructures found within complex networks, represented here primarily as nodes with specific input and output links. Embedded motifs are small, recognizable arrangements of nodes and edges that occur more frequently in a network than would be expected in a random or uniform distribution, precisely because they have played a causal historical role in forming and maintaining the network.

These patterns can be simple, such as three interconnected nodes within a level, or more complex configurations involving multiple nodes and connections across levels. The term *embedded* emphasizes a motif's existence within a larger network structure, where they are integrated into the network's topology. In the remaining chapters, we will discuss the individual causal learning cycle as a motif in team and organizational learning as well as cultural change, and we will bring forward the dynamics and states of this motif from level to level.

Studying these motifs as *structures of processes* provides insights into the organization, function, and dynamics of complex networks, revealing recurring structural motifs that play crucial roles in network behavior, functionality, and information processing. Individual learning within a team, for example, significantly influences the overall functioning and dynamics of both the team and the broader organization. Consider a scenario where a team comprising individuals with diverse expertise effectively collaborates on a project. As each team member engages in continuous learning and skill development, several impacts on the team and organization might become evident, including:

Enhanced Problem-Solving:

- **Skill Accumulation:** As individuals learn and acquire new skills, the team collectively gains a more diverse skill set, enabling multifaceted problem-solving approaches.
- **Innovative Solutions:** Individual learning fosters the introduction of fresh perspectives and innovative ideas, enhancing the team's capacity to devise creative solutions to complex challenges.

Improved Collaboration:

- **Knowledge Sharing:** Each member's learning journey brings new insights and knowledge, fostering a culture of knowledge sharing and peer-to-peer learning within the team.
- **Enhanced Communication:** As individuals learn and adapt, they refine their communication styles and abilities, facilitating clearer and more effective interactions within the team.

Organizational Impact:

- **Culture of Learning:** A team that values individual learning fosters an organizational culture centered on continuous improvement and adaptability.
- **Productivity and Adaptability:** Cumulative individual learning contributes to the organization's overall productivity and adaptability to evolving market trends and challenges.

Leadership Development:

- **Skill Transference:** The learning experiences of individuals can lead to the development of future leaders within the organization, as skills and insights are passed on through mentorship or leadership roles.

In summary, an individual's private learning cycle can act as a catalyst for team growth and organizational advancement in certain conditions. The accumulation of diverse skills, innovative problem-solving, improved collaboration, and the fostering of a collective learning culture can propel the team's effectiveness and contribute to the organization's resilience and success in an ever-evolving landscape. But this works only in certain conditions when team members are learning and contributing to the team effort. What decision-making role can AI play to help ensure the maximum effectiveness of each member? Are there homologous roles of a team that can catalyze change in a larger organization? Do knowledge communities then also potentially catalyze cultural shifts that impact more diverse and remote communities in the world at large?

We assert that a team's learning cycle embeds analogous network motifs from the micro into the meso level as *structures of learning processes*, that a similar transference occurs from the meso to the macro level, and that AI can learn the signs and signals of the processes to intervene at individual, team and large cultural levels. The individual interventions are described in Chaps. 4, 5, 6, and 7; the team interventions are described in Chaps. 8, 9, 10, and 11; the cultural interventions are described in Chaps. 12 and 13.

The term '*structures of processes*' forms a bridge between the concepts of static arrangements and dynamic activities, highlighting recurring configurations or arrangements that emerge and persist within the ongoing dynamics or temporal sequences of activities. Structures typically refer to the *static, stable, or enduring elements* within a system and are represented in the arrangement or organization of components, entities, or elements within a system at a specific point in time. For example, in network theory, structures could be specific network topologies, such as hubs, cliques, or hierarchical arrangements, which persist and evolve. In individual learning processes, structures include conceptual frameworks, schema, mental models, and aspects of metacognition (e.g. understandings of tasks, strategies, and self-knowledge). Processes, on the other hand, refer to *the dynamic, temporal, or ongoing activities, transformations, or sequences* within a system and represent the activities, changes, or transitions occurring over time within a system. For example, a process could be the flow of information, the evolution of a network, or the sequence of events in an individual problem-solving decision-making process. Blending these two ideas, *structures of processes* encompass the configurations, patterns, or frameworks that emerge and persist within an evolving or changing system. For example, in individual learning, maintaining and evolving one's schemas of problem-solving strategies is an example of such a structural process. As we intend the model to be clear about the physical mechanisms, we will think of all processes as *structures of processes*.

The three levels are arranged as a hierarchical system of nested networks with nodes and edges for computational modeling of learning processes. Figure 3.1 illustrates the 14 nodes of the network model: 6 nodes at the macro level (Engeström, 1999), 4 nodes at the meso level (Bransford et al., 2000), and 4 nodes at the micro level (Piaget, 1985). We briefly outline these nodes below, which later chapters further elaborate to include key relationships.

3.2.1 Individual Learning—The Micro Level

At the individual level, we associate two theories for the *states* of the learning cycle and another theory for the *motivating influences* linking the states, which are detailed more fully in Chaps. 4, 5, and 6 focusing respectively on Piaget, Keller, and Kauffman. In those chapters, we outline several historical theories of learning and describe how the use of the Kauffman mechanism expands understanding of the mechanisms involved in the various theories. In brief, where the historical theories are observational and functional projections based on learning behaviors, the Kauffman mechanism is a physical theory underlying a new concept of *autocatalysis of agency for learning*. The Kauffman mechanism provides detailed physical descriptions for concepts from the historical theories of the psychology of learning.

These elements are integrated to form a ubiquitous engine of learning, the *causal learning cycle*, with Piagetian *states or phases* of the processes involved in cognition: disequilibration, assimilation, accommodation, and equilibration (Piaget, 1954, 1985) (Fig. 3.2).

The four Piagetian phases are often associated with the adaptation of schemas or mental frameworks to new information:

1. **Disequilibration** refers to a state of cognitive imbalance or discomfort experienced by individuals when they encounter new information or experiences that do not align with their existing mental schemas or understanding of the world.
2. **Assimilation** involves the process of incorporating new information or experiences into existing mental frameworks or schemas without altering those frameworks significantly. It's when individuals interpret new information based on their current understanding.
3. **Accommodation** is the process of modifying or adapting existing mental frameworks or schemas to incorporate new information or experiences that cannot be assimilated. It involves altering one's cognitive structures to better fit or accommodate the new information.

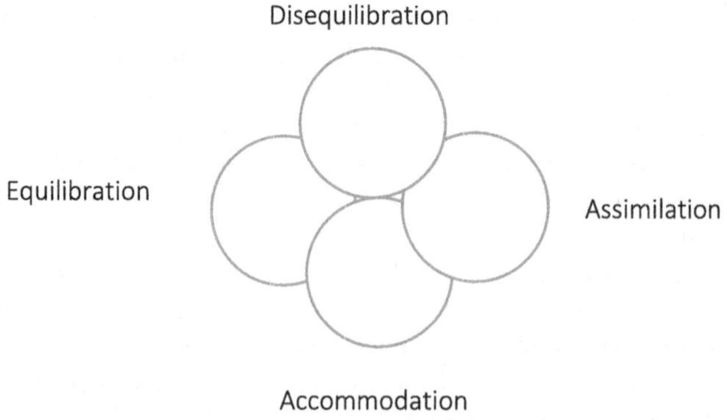

Fig. 3.2 Phases in Piaget's equilibration theory—processes of adaptation

4. **Equilibration** refers to the process of achieving a state of balance or equilibrium between assimilation and accommodation. It involves resolving cognitive conflicts by adjusting mental schemas to accommodate new information, leading to a more comprehensive and balanced understanding of the world.

The four phases represent the dynamic processes through which individuals adapt and refine their cognitive structures to accommodate new experiences, fostering cognitive growth and development.

The influences on the edges of the network graph of the system are defined in Chap. 5 with Keller *ARCS relationships or dynamics* (Keller, 2009; Song & Keller, 1999) labeling the transformations (i.e., attention, relevance, confidence, and satisfaction). The key concepts of the ARCS model are used to label the outcomes of transitioning from one Piagetian node to another, a process that is understood to be an imperative latent driver of sustainability, reproduction, learning, and growth during evolution. Figure 3.3 shows how *attention* mediates and influences the phase from disequilibrium to assimilation, *relevance* influences accommodation, *confidence* influences equilibration and *satisfaction* influences disequilibration.

In Chap. 6 we add a new interpretation to the cognitive phases based on Stuart Kauffman's foundation for computational biology's explanation of *autopoiesis*, or self-organization (Kauffman, 2000) which allows us to discuss the causal learning cycle as the transformation of energy into structure, information and action. Autocatalytic agency, which proposes an understanding of how complex living systems self-organize and exhibit emergent properties without the need for external direction or control, is appropriated to account for self-motivated learning and the instinct to continuously explore and make sense of the world at the micro level. At the meso level, the concept of autocatalytic agency is appropriated to account for

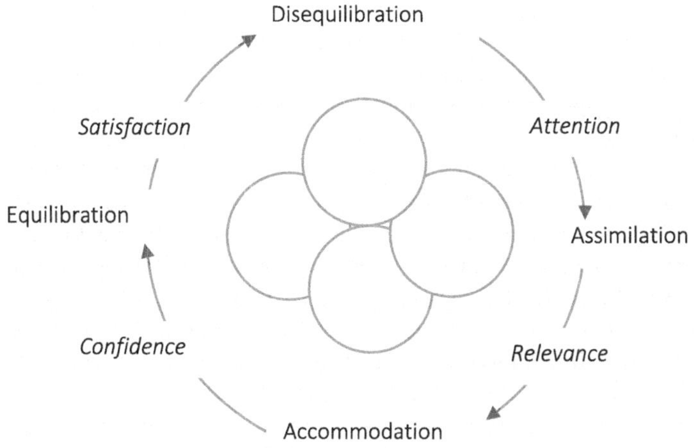

Fig. 3.3 Micro level nodes and edges. Note: Two theories are combined to define the primary nodes and edges of the micro-level network. Equilibration (Piaget, 1985), and Motivation (Song & Keller, 1999)

team formation and knowledge community evolution, and at the macro level, for the emergence of new fields and cultural evolution.

The additional Kauffman dimension of the causal learning cycle pinpoints a physical mechanism for constructivism, the psychological theory of the construction of knowledge. The core idea is that cognition utilizes energy to produce ideas and action (e.g. food is ingested, and its energy is transformed by the work of building bodies and minds) and can therefore be understood further with the science of thermodynamics, which provides a physical model for the transformation of energy into structures and processes such as memories, skeletons, body systems, schemas, and functional capabilities of the mind. Autocatalysis, according to Kauffman, takes place via a four-part idealized cycle of work. Potential energy from outside the agent is ingested causing disequilibrium, the energy is then turned into kinetic energy for building internal structures and memories of processes, then materials and information are passed outside the agent as work products or waste, and the engine returns to a state of dynamic equilibrium, perhaps expanded, more elaborated and more intelligent.

The steps of the four-phase Kauffman process—*isothermal compression and expansion* and *adiabatic compression and expansion*—can be aligned with Piaget's insight and by doing so adds physical detail concerning how cognition builds knowledge. We conjecture that the compressibility of natural phenomena that gives rise to meaning in symbolization (see for example (Baum, 2003; Chaitin, 1975; Langer, 1953)) may utilize both iso and adiabatic compression. We discuss this in more detail in Chap. 6.

The 3-level model has now arrived at a new understanding of Piaget's insight into adaptation. Disequilibrium physically pushes cognition to (isothermally) expand followed by mental model (adiabatic) expansion by assimilation. The act of symbolization during accommodation takes physical energy to actively (isothermally) compress the regularities of nature into a shorter semantic 'handle' for mental manipulation, and the system returns to near-normal equilibrium as an all-internal (adiabatic) process of resettling into a state of readiness for responsivity (Fig. 3.4). The physical and psychological ramifications are discussed in the section on individual learning.

The micro-level model thus explains some of the key internal states and processes of individual cognition. However, also in need of explanation are the outcomes of cognitive processes, for example, internal structures and representations such as mental models, schemas, memories, automatic skill complexes, habits, and so forth (Ifenthaler, 2011; Ifenthaler & Seel, 2013) and externalizations such as traces from actions, things made and said, as well as tools and symbols used during problem-solving. In addition, inputs to these processes are provided by both the macro and meso levels, which are discussed in later sections.

The micro level as outlined above contains a *causal network model of individual learning processes* that provides a framework for understanding how AI can be integrated to support individual learning. Given these new insights into the core of individual learning processes, we share conclusions concerning AI roles at the micro level in Chap. 7.

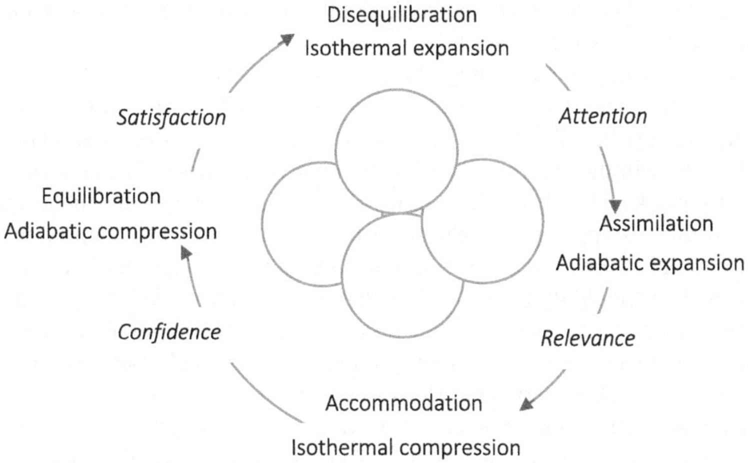

Fig. 3.4 Piaget-Keller-Kauffman (PKK) mechanisms of cognition. Note: According to (Kauffman, 2000) idealized thermodynamic concepts are a foundation for understanding autocatalysis and the emergence of life and therefore intelligent life

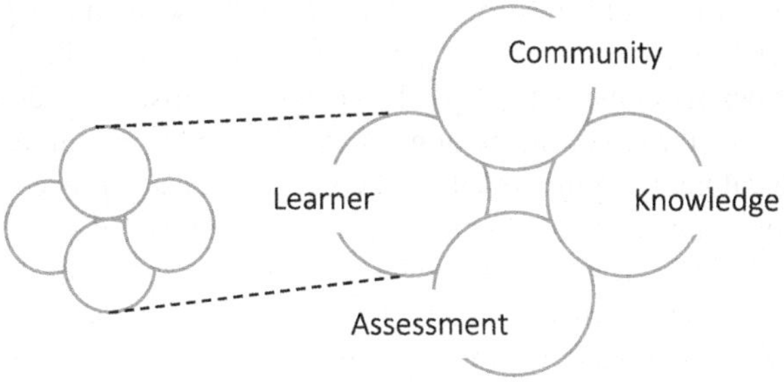

Fig. 3.5 Meso level—HPL (How People Learn) nodes. Note: The four components of the HPL model of the meso level include the Community of practice, its specialized Knowledge and ways of knowing, its members including Learners who aspire to expertise in the community, and its processes of Assessment and validation. Individual members from the micro level are replicated at the meso level

3.2.2 Social Learning—The Meso Level

At the meso level is the team, task group, or knowledge community, in which the micro-level mechanisms (i.e. the states and dynamics of the causal learning cycle) are combined with social network theory and social psychology to render what John Dewey noted as a *communal process of inquiry* (Dewey, 1916) which we elaborate in Chap. 8. Donovan, Bransford, Pellegrino, and others have explained this level as How People Learn (HPL) in the *social context of a learning community* (Bransford et al., 2000; Donovan et al., 1999; Pellegrino et al., 2001), illustrated with four components in Fig. 3.5 which we discuss in Chap. 9. With the introduction of computers in education, this level has also been described as a *critical community of inquiry enhanced by technology* (Garrison et al., 2009) which we will extend to include AI enhancements in Chaps. 10 and 11.

In social learning, a community of practice is motivated to stimulate new knowledge, tools, and processes and attract new members. It is most often researched from the standpoint of public discourse, which we conjecture may play a role homologous with the individual's self-talk at the micro level. Evidence of a functional crosswalk and embedding from the private micro level to the public space of the meso level is apparent, for example, in the descriptors of cognitive presence (Garrison et al., 2009): *perception* and *awareness* maps to disequilibration, *deliberation* and *applicability* maps to assimilation, *conception,* and *ideation* maps to accommodation, and *experience* and *practice* maps to equilibration by bridging from ideation to its realization and satisfaction. In this example, we see homologous functions of the causal learning cycle at a new level.

Where the *learner* is the primary agent at the micro level, the *expert community* is the primary agent at the meso level. The community provides induction that inculcates new members in the community's knowledge and ways of knowing (Collins et al., 1989; Csikszentmihalyi, 1996). Formal professional feedback by the community is ideally guided by evidence-centered models (Mislevy et al., 1999) that organize the knowledge domain, characteristic tasks, and expected capabilities of practitioners in the field. Informal forms of feedback include signals of social acceptance, pats on the back, critical questions involved in collaborative inquiry, and so forth. Socially constructed knowledge and expertise are thereby both replicating and evolving through contributions by existing and new community members.

The dynamical model at the social level contains state and process combinations formed into dynamic clusters of embedded motifs composed of dissipations and aggregations from the micro level as well as emergent relationships within the meso level. This is discussed further in Chaps. 10 and 11. Dissipation (e.g. active uses of energy to get work done) and aggregation processes (e.g. passive uses of energy to store, access, and juxtapose work products) are elaborated further in Chap. 6 which includes a discussion of complexity processes in learning science. The states and processes projecting into the meso level from the micro level are viewed as latent variables, emergent roles, and underlying driving mechanisms for meso-level processes. For example, a single learner at the micro level becomes a member of a team or group at the meso level and the current state of one learner's PKK interacts with another person's PKK as well as team dynamics, mission, and assessed status.

A novice who is not yet a member of the expert community might introduce a new idea that causes disequilibration in the community, and the community must assess whether to assimilate the new person and knowledge. If the community opts to do so, then the new knowledge and person must be accommodated, and when that has been accomplished, equilibration returns. Thomas Kuhn (1962) noted this level's entities and processes as the core of scientific revolutions, and Mihaly Csikszentmihalyi (1996) described the cross-level entities and relationships from individual, to social, to cultural systems, which we will call upon in Chaps. 8 and 9 to help shape the dynamics within the meso social level and indicate how level-crossing influences may work.

3.2.3 Sociocultural Evolution—The Macro Level

At the macro level, we introduce the concept of 'group of groups' by which we mean to encompass cross-disciplinary activity, international activity, and broadly, the level of complexity where culture is influential and evolves. At this level, the Cultural Historical Activity Theory (Engeström, 1999) is used as a 6-node network model for understanding the larger sociocultural systems involved in learning (Fig. 3.6). We elaborate further in Chap. 12.

The 4-node meso-level network (i.e. the HPL model) participates as one of several communities at the macro level and exerts partial influence in clusters of the 6-node macro level's entities (see below for a brief introduction to the mechanisms of *polysemantic roles* and *persistent homologies*). For example, Artifacts, Objects, and Rules at the macro level are unified by an expert community to become aspects of the community's Knowledge at the meso level, described further in Chaps. 12 and 13. At the same time, macro level Community, Role, and Subject combine to partially constitute Feedback processes at the meso level, for example, to inculcate a novice into becoming an expert. At the macro level, the constituents influence the development of new fields of knowledge in other communities as community members take on roles outside of their field of expertise, such as a poet becoming a politician, an engineer becoming a musician or a musician becoming a learning theorist.

Depending on the context of an inquiry, a Subject at the macro level may be an individual from the micro level (e.g. a member of the international SOLAR community) who is a member of an HPL group of learners from the meso level (e.g. the editors of a SOLAR journal), an existing community (e.g. the SOLAR organization) or an emergent new interdisciplinary 'group of groups,' such as data scientists and learning scientists working in the domain of educational technology. The SOLAR organization is an interdisciplinary network of leading international researchers who are exploring the role and impact of analytics on teaching, learning, training, and development.

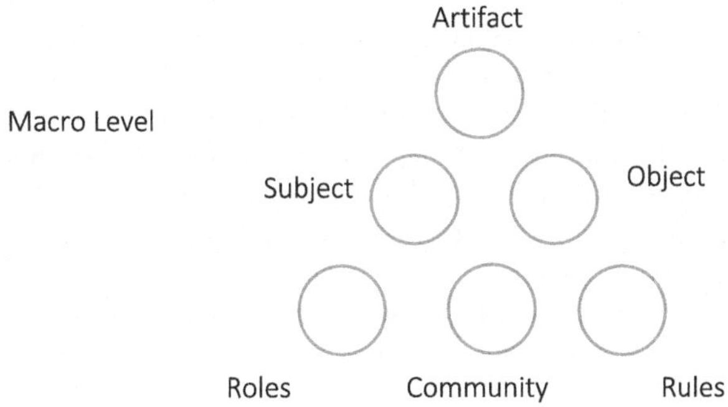

Fig. 3.6 Cultural historical activity theory model of the macro level. Note: The HPL model of a meso-level expert community, itself containing replications of the micro-level PKK model of individual learning, is replicated at the macro level into the node labeled "Community"

In each case of analysis with the macro lens (e.g. whether considering a micro, meso, or macro phenomenon), the definition of 'community' changes in each to ensure that the concept represents an encompassing larger context containing the 'unit of analysis' as a member. We elaborate on cultural influences and draw links to the potential AI roles that enhance cultural change in Chap. 13.

3.3 Dynamic Causal Network Concepts in the Trilevel Model

Next, we will discuss the meaning of a few key terms needed to integrate complexity and machine learning into learning science, such as *dynamic, causal,* and *network,* and then bring the terms together as the key characteristics of individual learning processes, team and group learning processes, and cultural change in the tri-level model.

In an earlier era when computers were new in education, Seymour Papert envisioned computers and programming languages as *mediating tools* that could change how children think, solve problems, and construct knowledge (Papert, 1980). With AI as an embedded cognitive partner, understanding and modeling the mediating power of AI to promote learning processes in individuals, groups, and cultures is hinged on *complexity* and *machine learning.*

Complexity theory, which we view as a formal expansion of the theory of evolution into computational modeling, emphasizes the interactions and feedback loops that constantly drive change in open, dynamic systems. While complexity theory proposes that such systems are at one level unpredictable, they are also constrained by order-generating rules within some bounded infinity that separates the system from the rest of the universe. This view accords for example, with algorithmic uncertainty and bounded infinity as discussed by (Chaitin, 1975, 2003), and generative algorithms as discussed by (Wolfram, 2002) both of which build on mathematical ideas of Leibnitz, Cantor, Hilbert and others. The core idea of genetic algorithms (Holland, 1998; Holland & Reitman, 1998) which we conjecture is a computationally well-defined physical mechanism for editing and adapting mental models, schemas, and the like, leads naturally to the processes of induction, abduction, and deduction (Holland et al., 1986; Thagard, 2010) that are key to understanding cognition. We discussed complexity in Chap. 1 as part of the rationale for asserting that the learning sciences from this point forward need to be *computational* to leverage the most from AI. Despite the uncertainties and complications, complex systems concepts, as we hope to elaborate, are foundational for explaining and probabilistically predicting the adaptive capabilities embedded at every level of hierarchical evolutionary systems of learning unfolding over time, within and across the micro-meso-macro levels.

Machine learning, in the context being established here for computational learning science, is understood as an operational approach to researching learning models, in addition to its usual definition as a method for building computer systems that learn from data. Machine learning algorithms are usually trained to find relationships and patterns in data; they use historical data as input to make predictions, classify information, cluster data points, and reduce dimensionality. Since the advent of large language models, machine learning approaches now also profusely and creatively generate new content. Machine learning concepts important to AI discussed in this book include:

Transfer learning enables models to leverage knowledge from one domain or task and apply it to another, facilitating quicker learning in new scenarios and domains.

Meta-learning, in which algorithms adapt and improve their learning strategies based on previous experiences, aids in rapid adaptation to new tasks.

Reinforcement learning, teaches computational agents to make sequences of decisions to maximize rewards in various environments, enabling AI systems to learn through interaction and decision-making.

Generative models that enable AI systems to generate new content, fostering creativity and imagination in problem-solving and content creation.

Self-supervised learning, allows models to learn from unlabeled data by creating tasks from the data itself, enabling AI to learn rich representations and understand context.

Attention mechanisms that enable models to focus on relevant parts of input data, improving contextual understanding and facilitating more human-like reasoning.

Explainable AI methods that aim to make AI systems' decision-making processes transparent and understandable, fostering trust and facilitating human-machine collaboration.

Continual learning techniques allow AI systems to accumulate knowledge over time, preventing catastrophic forgetting and enabling continuous adaptation to new information.

In the chapters to come, the relationships of the above machine learning ideas to notions and theories in the psychology of learning will be discussed to underpin our assertion that to leverage the most from AI, learning science theories from this point forward must have computational as well as psychological, social and cultural realizations.

Machine learning concepts important to AI are strongly aligned in the proposed 3-level model, we assert, via persistent homologous functional roles (Aktas et al., 2019) that help explain time-tested learning theory models and these functional roles should be considered in existence proofs, and validations of the 3-level model. *Persistent homology*, according to Aktas, is a mathematical tool in computational topology that measures the topological features of data that persist across multiple scales, so is an apt concept for helping traverse from micro, to meso, to macro levels of the learning processes system. A homologous or similar functional role that is found in two or more levels, we propose, is a topological feature of the data space of the modeling environment. An example of such a role is the 'learner' (i.e. the Piaget-Kauffman-Song-Keller *causal learning cycle)* at the micro level, which becomes a 'learning team' (i.e. the Bransford et al. *social learning network)* at the meso level and a 'learning community' (i.e. the Engestrom et al. *cultural historical learning network)* at the macro level. In each level, similar learning processes exist with functions roughly equivalent to disequilibration causing assimilation causing accommodation causing equilibration (e.g. of the individual's equilibrium cognitive state, the team's collective intelligence state, and the culture's current state of norms).

By *dynamic* we mean a mechanism that advances a process over time (recall Eq. 1.1 in Chap. 1), usually representable by an algorithm or pseudocode that uses simple language to describe an entity node or process edge. A network *dynamic* advances the state of the network by instigating changes in the relationships, interactions, or configurations among its constituent elements, thereby altering the network's structure, functionality, or emergent properties. In learning science, the dynamic mechanism we are referring to as the causal learning cycle is a natural system of parts and processes that initiates and mediates individual and organizational growth and change in some combination of behavior, emotion, and cognition at the individual level. At the meso and macro levels, we find appropriate pre-adaptations of analogous dynamics for social settings like teams and organizations, and for larger cultural entities we find them in the arts and sciences, society at large and world culture.

The word *causal* implies that any point of observation or analysis, regardless of where a researcher observes a phenomenon, probabilistically *entails,* that is, the observation strongly implies and at least probabilistically necessitates a predicted next step within the level and if the level's process or stage influences another level, that too entails a probable 'next step' influence within the new level.

A *dynamic causal network* then, is a particular kind of network model (Sugihara et al., 2012), originally developed for testing hypotheses about neural dynamics, which consists of nodes representing states and edges representing causal influences. As the system (i.e., learner, group, or culture) evolves, the *state* of the system refers to the simultaneous status of all relevant parts or components, and the *dynamics* or processes of the system refer to the relationships, transitions or causal influences from part to part and state to state.

Separation between the levels in the tri-level model is defined by both inclusion and exclusion criteria. States and processes that belong solely to individual learning belong at the 'micro.' States and processes that can only be introduced if there are two or more people involved in learning together belong at the 'meso' and states and processes that involve two or more learning communities, or groups of subject area experts, exist at the 'macro' level.

Permeable bivalent boundaries for entities within each level allow nodes to act in their causal role or as mechanisms in one context and states in another, and this bivalent capability supports participation in both composition and influence. The machine learning literature has explored the *superposition* of polysemantic nodes (Elhage et al., 2022) and models (Cheung et al., 2019) such as node or cell assemblies (Zenke et al., 2015) as network mechanisms for multiple encodings. In neuroscience, a cell assembly is loosely formulated as a group of neurons with strong connections among each other interpreted as a functional circuit of brain activity. The tri-level dynamical hierarchy theory envisions making use of superposition to unfold an appropriate sub-model or polysemantic node within a specific context, similar, for example, to the way prompts given to a large language model shape a context for appropriate vector selection and interpretation of inputs during a process such as a 'chain of thought' (Wei et al., 2022). When a context demands it, semantic shifts to another embedded model become more probable. Another related example is the 'free energy' framework for modeling perception (Bogacz, 2017) in which synaptic plasticity only depends on the activity of pre-synaptic and post-synaptic neurons (as opposed to an over-riding model of the world or encompassing context), regardless of where the node sits in a hierarchy. Professor Bogacz and colleagues have called the new contextual learning principle 'perspective configuration' (Song et al., 2022). These concepts will be discussed in each section of the model as they form a basis for reflectivity and bounding the infinite within each and between levels.

Further, to add specificity concerning the connections between the layers, the model needs to reinterpret or *homologize* the mechanisms from the individual learning cycle terms into terms more commonly understood from collaborative learning (meso) and cultural change (macro). For example, 'disequilibration' in an individual may be understood as "anomaly and crisis" at the meso level (Kuhn, 1962) and "second-order cybernetics" (Bateson & Bateson, 1972; Von Foerster, 1979) at the macro level. This is an example of what we mean by 'homologous functions' in the three levels, which we will include in all relevant chapters. To bring the functional outlook into focus for educational research, a potential example of perspective configuration at the cellular level, when extended into the cultural level, is illustrated by the idea of 'prospective cognition' (Critchfield & Jordan, 2014; Zimmerman & Croker, 2014).

Finally, the causal linkage requirements of the theory present clear and testable hypotheses needed for developing computational models and other forms of research for further study, which we elaborate on in Chaps. 14 and 15.

3.4 Limitations and Conclusions

This chapter did not attempt to fully address how to practically implement AI in all education settings, focusing instead on 'learning processes' at micro, meso, and macro levels. It did not consider many of the diverse learning characteristics and needs of individual learners in specific detail, but instead aggregated those into a larger concept of the 'learner.' In addition, the complexity of the 3-level hierarchical framework may be difficult for some to understand, which hopefully subsequent chapters will help explain. The chapter also did not cover many of the limitations of AI, such as biases in machine learning algorithms, nor discuss the ethical implications of using AI in several educational contexts, reserving those discussions for later in the narrative.

The chapter asserted that AI applications can support individual learning by integrating with a causal network model of learning processes and that learning science theories should now include computational, psychological, social, and cultural aspects in the computational modeling of learning processes. Cultural learning is a potential context for social learning and social learning is a potential context for individual learning.

Summing up, the 3-level model integrates learning theories into a single computational framework. Micro-level learning is the private universe of individual attention that sometimes leads to the recognition of information that either fits into existing knowledge or compels the formation of new awareness that in turn creates new mental models and new levels of responsiveness to a surrounding context. Meso-level learning involves learning within a group or team that possesses a shared base of knowledge (e.g. possesses shared problems, reaches shared conclusions), including how to assess what members know and can do, and that shapes its members' expertise via feedback, which often requires the individual to face and integrate different perspectives. Macro-level learning involves interdisciplinary processes like knowledge transfer and expert community integration and can be analyzed as a cultural-historical activity.

3.5 Moving Forward

We are now ready to embark on a journey into the tri-level model hierarchy and its encompassed dynamic relationships to address questions such as: How can AI assist individuals in their learning processes? How can AI assist expert teams, collaborative teams and knowledge communities in their learning processes? How can AI assist a larger interdisciplinary culture in its learning processes?

References

Aktas, M. E., Akbas, E., & Fatmaoui, A. E. (2019). Persistence homology of networks: Methods and applications. *Applied Network Science, 4*(1). https://doi.org/10.1007/s41109-019-0179-3

Bateson, G., & Bateson, G. (1972). *Steps to an ecology of mind: Collected essays in anthropology, psychiatry, evolution, and epistemology*. Chandler Pub. Co.

Baum, E. (2003). *What is thought*. MIT.

Bogacz, R. (2017). A tutorial on the free-energy framework for modelling perception and learning. *Journal of Mathematical Psychology, 76*, 198–211. https://doi.org/10.1016/j.jmp.2015.11.003

Bransford, J. D., Brown, A. L., & Cocking, R. R. (2000). *How people learn: Brain, mind, experience, and school*. http://www.nap.edu/openbook.php?isbn=0309070368

Chaitin, G. (1975). A theory of program size formally identical to information theory. *Journal of the ACM, 22*(3), 329–340. https://doi.org/10.1145/321892.321894

Chaitin, G. (2003). Algorithmic information theory. In *Cambridge tracts in theoretical computer science*. Cambridge Univesity Press.

Cheung, B., Terekhov, A., Chen, Y., Agrawal, P., & Olshausen, B. (2019). Superposition of many models into one. *ArXiv*. http://arxiv.org/abs/1902.05522

Collins, A., Brown, J., & Newman, S. (1989). Cognitive apprenticeship: Teaching the crafts of reading, writing, and mathematics. In *Knowing, learning, and instruction: Essays in honor of Robert Glaser* (pp. 363–401). MacMillian.

Critchfield, T., & Jordan, J. (2014). Prospective cognition in education and enculturation: An overview. *Journal of Cognitive Education and Psychology, 13*(2), 139–146. https://doi.org/10.1891/1945-8959.13.2.139

Csikszentmihalyi, M. (1996). *Creativity: Flow and the psychology of discovery and invention*. HarperCollins.

Dai, W., Lin, J., Jin, F., Li, T., Tsai, Y.-S., Gasevic, D., & Chen, G. (2023). *Can large language models provide feedback to students? A case study on ChatGPT* [Preprint]. EdArXiv. https://doi.org/10.35542/osf.io/hcgzj

Dewey, J. (1916). *Democracy and education: An introduction to the philosophy of education*. Macmillan.

Donovan, S., Bransford, J., & Pellegrino, J. (1999). How people learn: Bridging research and practice. (N. R. Council, Ed.). National Academy Press.

Elhage, N., Hume, T., Olsson, C., & Schiefer, N. (2022). *Toy models of superposition*. Transformer Circuits Thread.

Engeström, Y. (1999). Activity theory and individual and social transformation. In Y. Engestrom, R. Miettinen, & R. Punamaki (Eds.), *Perspectives on activity theory* (pp. 19–38). Cambridge University Press.

Garrison, D. R., Anderson, T., & Archer, W. (2009). Critical thinking, cognitive presence, and computer conferencing in distance education. *American Journal of Distance Education, 21*(1), 7–23. https://doi.org/10.1080/08923640109527071

Giordano, C., Brennan, M., Mohamed, B., Rashidi, P., Modave, F., & Tighe, P. (2021). Accessing artificial intelligence for clinical decision-making. *Frontiers in Digital Health, 3*, 645232. https://doi.org/10.3389/fdgth.2021.645232

Holland, J. (1998). *Emergence: From chaos to order*. Helix Books Addison Wesley/Perseus Books Group.

Holland, J., & Reitman, J. (1998). Cognitive systems based on adaptive algorithms. In *Evolutionary computation: The fossil record*. https://doi.org/10.1109/9780470544600.ch16

Holland, J., Holyoak, K., Nisbett, R., & Thagard, P. (1986). *Induction: Processes of inference, learning, and discovery*. MIT Press.

Huang, J., Gu, S. S., Hou, L., Wu, Y., Wang, X., Yu, H., & Han, J. (2022). *Large language models can self-improve*. http://arxiv.org/abs/2210.11610

Ifenthaler, D. (2011). Bridging the gap between expert-novice differences: The model-based feedback approach. *Journal of Research on Technology in Education, 43*(2), 103–117.

Ifenthaler, D., & Seel, N. M. (2013). Model-based reasoning. *Computers & Education, 64*, 131–142. https://doi.org/10.1016/j.compedu.2012.11.014

Jones, K. (2023, December 3). *Nine problems with generative AI*. https://www.visualcapitalist.com/sp/9-problems-with-generative-ai-in-one-chart/

Kauffman, S. (2000). *Investigations*. Oxford University Press.

Keller, J. M. (2009). *Motivational design for learning and performance: The ARCS model approach* (1st ed.). Springer.

Kuhn, T. (1962). *The structure of scientific revolutions*. University of Chicago Press.

Kumari, N., & Pandey, S. (2023). Application of artificial intelligence in environmental sustainability and climate change. In *Visualization techniques for climate change with machine learning and artificial intelligence* (pp. 293–316). Elsevier. https://doi.org/10.1016/B978-0-323-99714-0.00018-2

Langer, S. K. (1953). *Feeling and form*. Charles Scribner's Sons.

Mislevy, R., Steinberg, L., & Almond, R. (1999). *Evidence-centered assessment design*. http://www.education.umd.edu/EDMS/mislevy/papers/ECD_overview.html

Nature Editors. (2023, December 3). *Nature Portfolios, Artificial Intelligence*. https://www.nature.com/nature-portfolio/editorial-policies/ai

Papert, S. (1980). *Mindstorms: Children, computers and powerful ideas*. Basic Books.

Pellegrino, J. W., Chudowsky, N., Glaser, A. R., & Glaser, R. (2001). Knowing what students know: The science and design of educational assessment. In *Committee on the foundations of assessment, board on testing and assessment*. Center for Education, National Research Council (Vol. 24). National Academy Press. https://doi.org/10.1037/e378022004-001

Piaget, J. (1954). *The construction of reality in the child* (M. Cook, Trans.). Basic Books. https://doi.org/10.1037/11168-000

Piaget, J. (1985). *The equilibration of cognitive structures: The central problem of intellectual development*. University of Chicago Press.

Song, S., & Keller, J. (1999). *The ARCS model for developing motivationally-adaptive computer-assisted instruction*. National Convention of the Association for Educational Communication and Technology.

Song, Y., Millidge, B., Salvatori, T., Lukasiewicz, T., Xu, Z., & Bogacz, R. (2022). *Inferring neural activity before plasticity: A Foundation for learning beyond backpropagation* [Preprint]. Neuroscience. https://doi.org/10.1101/2022.05.17.492325

Sugihara, G., May, R., Ye, H., Hsieh, C., Deyle, E., Fogarty, M., & Munch, S. (2012). Detecting causality in complex ecosystems. *Science New York, N.Y., 338*(6106), 496–500. https://doi.org/10.1126/science.1227079

Thagard, P. (2010). How brains make mental models. In L. Magnani, W. Carnielli, & C. Pizzi (Eds.), *Model-based reasoning in science and technology: Abduction, logic, and computational discovery* (pp. 447–461). Springer. https://doi.org/10.1007/978-3-642-15223-8_25

Von Foerster, H. (1979). Cybernetics of cybernetics. In K. Krippendorff (Ed.), *Communication and control in society*. Gordon and Breach.

Wei, J., Wang, X., Schuurmans, D., Bosma, M., Ichter, B., Xia, F., Chi, E., Le, Q., & Zhou, D. (2022). Chain of thought prompting elicits reasoning in large language models. *arXiv*. http://arxiv.org/abs/2201.11903

Wolfram, S. (2002). *A new kind of science*. Wolfram Media.

Zenke, F., Agnes, E. J., & Gerstner, W. (2015). Diverse synaptic plasticity mechanisms orchestrated to form and retrieve memories in spiking neural networks. *Nature Communications, 6*(1), 6922. https://doi.org/10.1038/ncomms7922

Zhou, Y., Muresanu, A. I., Han, Z., Paster, K., Pitis, S., Chan, H., & Ba, J. (2022). Large language models are human-level prompt engineers. *arXiv*. http://arxiv.org/abs/2211.01910

Zimmerman, C., & Croker, S. (2014). A prospective cognition analysis of scientific thinking and the implications for teaching and learning science. *Journal of Cognitive Education and Psychology, 13*(2), 245–257. https://doi.org/10.1891/1945-8959.13.2.245

Chapter 4
Piaget and the Ontogeny of Intelligence

Intelligence, the most plastic and at the same time the most durable structural equilibrium of behavior, is essentially a system of living and acting operations.

—Jean Piaget

4.1 Introduction

Jean Piaget (1896–1980) was a pioneering Swiss psychologist, renowned for his groundbreaking work in developmental psychology and the study of cognitive development in children. Born in Neuchâtel, Switzerland, Piaget demonstrated an early interest in biology and the natural sciences. He later pursued studies in psychology, ultimately becoming one of the most influential figures in developmental psychology. Piaget's seminal contributions revolved around his theory of cognitive development, proposing that children progress through distinct stages of intellectual growth, each characterized by unique ways of understanding the world. His observations and experimental research highlighted the importance of children's active exploration and interaction with their environment in shaping their cognitive development. This book's narrative will not have more to say about the developmental aspect of his theory, which has been extensively covered in the psychological literature e.g. (Bruner, 1997), but instead focus on his *theory of equilibration,* which Piaget proposed as the mechanism that drove the progressive development of intelligence in children.

In his early twenties around 1919, Piaget worked as an assistant to Alfred Binet, who created the first official IQ test, administrating intelligence tests to children. During this time, he became fascinated by the children's responses to the tests, noticing distinct patterns in the errors made by younger children compared to older ones. These observations sparked Piaget's curiosity and prompted him to delve deeper into understanding the cognitive development of children, the groundwork for his research on cognitive development and children's intellectual growth.

© The Author(s), under exclusive license to Springer Nature Switzerland AG 2024

D. C. Gibson, D. Ifenthaler, *Computational Learning Theories*, Advances in Analytics for Learning and Teaching,
https://doi.org/10.1007/978-3-031-65898-3_4

Piaget's work significantly influenced educational practices and remains a cornerstone in understanding how children acquire knowledge and construct their understanding of the world. Here we focus on Piaget's theory of the core mechanism of the ontogeny of intelligence, beginning with the child's conception of time.

In *The Child's Conception of Time* (Piaget, 1946) he explained how children conceptualize and grasp an abstract concept via stages of cognitive development of temporal understanding. Children move from a lack of differentiation between past, present, and future to developing more structured concepts of duration, temporal order, and the relationship between time and events. Children initially view time through their personal experiences before developing a more objective understanding influenced by social interactions, and they gradually acquire the ability to take different temporal perspectives, understanding the time-related viewpoints of others, and how these perspectives influence their comprehension of time. The cognitive progression from subjective, egocentric perceptions towards more objective, socially influenced understandings presages an aspect of our proposed scalable, adaptable learning theory framework, namely that a process of grasping reality (e.g. by individuals, teams and cultures) can be found at each level, functioning in similar ways to drive learning and growth.

4.2 Cognitive Equilibration Theory

Supporting the conceptualization of cognition progressing from private to public spaces during maturation, the theory of *cognitive equilibration* (Piaget, 1985) explains how new information is integrated with existing knowledge. Coming to know something involves processes of assimilation or fitting new information into existing mental schemas and accommodation, adjusting or changing a schema to fit new information. The process begins with disequilibration caused by noticing something different in the world and progresses through a standard sequence, a causal chain. We redefine and elaborate on the four phases, place them in a causal order, and discuss them as the primary mechanism of learning—a causal learning cycle—and then propose that the cycle is a feature found at every level of the complex system of individual, social, and cultural intelligence.

A modern reflection on the functions envisioned by Piaget provokes a search for a combined computationally feasible physical and neurological mechanism that can integrate into the four phases. This new understanding has awaited the development of high-performance computing machines, which have aided the understanding of the science of irreversible dynamics and complexity. The new understanding is relevant to our discussion about AI enhancements of learning processes because until computational tools made the concepts accessible to anyone, understanding and developing intuitions about complex dynamics was limited to a few mathematicians and theorists. While the child in Piaget's observational terms developed a 'naïve model' of causation and time, a modern understanding includes the concept that processes in the real world are also *irreversible*. This understanding departs from classical physics and needs to be explained, according to (Drossel, 2023).

Briefly, *irreversibility* means that physical processes in the real world cannot be undone. Real events, as opposed to ideal laboratory experiments, create a unique 'frozen history' of artifacts that can never return to the initial condition. A glass that falls from the shelf and breaks will not spontaneously reassemble itself, although much of classical physics assumed that it possibly could. During an irreversible process, *entropy* or noise increases. The mystery of life is that it fights uphill against entropy (*antientropic)* to create structure and maintain itself against the thermodynamics of the universe's winding down to heat death. Recall that in Chap. 1 the noisiest chaotic condition of the dynamic equation resulted in an ever-novel sequence of events, but under different starting conditions, could also propel a simple sequence of events or even a single definitive ending point. These 'edge of chaos' system evolutions will be discussed at greater length in Chap. 6, where irreversibility, history, and chaotic regimes of complex systems combine to create the universal motivation to 'learn forward' into the unknown.

How might irreversibility relate to learning processes? The Piagetian return to *equilibrium* is not a return to the exact starting condition nor to a static condition (as the term implies in classical thermodynamics), but now includes learning and growth that has been added either consciously or unconsciously during assimilation and accommodation. *Equilibration* is a momentary, dynamic phase of the learning process over time. The four Piagetian phases are not static stages, but dynamic moments in an antientropic uphill climb to survive, persist, grow, and change. The reinterpreted Piagetian framework as elaborated now includes a physical theory of the transformation of experience into a variety of structures of intelligence (e.g. embodied intelligence, memories, simple heuristics, expert skills, and creative potential).

The transformation employs a series of mechanisms that store and then utilize energy and information that expand the agent's psychological and physical capabilities into the *adjacent possibility space* (Kauffman, 2000) of experience of the learner, team, or culture. Piaget had insight into this system and expressed it in terms familiar to the science of his day, however, now we can appreciate and acknowledge his insights while adding new layers of meaning and sources of insight via the 'charity of interpretation' (Putnam, 1992). Such leeway via charity of meaning, in theory, allows us to acknowledge, for example, that the early Greeks invented the word 'atom' and pointed to essentially the same phenomena we mean by the term today, even though they knew nothing about the details of quarks, leptons and the unification of the laws of nature in quantum theory.

Likewise, Piaget did not have readily at hand, knowledge of the computational universe, how order can arise from chaos, and the entwining of entropy with life, as has recently been found to measure the possibility of life in other parts of the universe (Petraccone, 2023). So, the process of elaboration on Piaget will begin with the concepts of *time, irreversibility, and entropy as critical to self-organization*, as these are signs of life and where there is life, there is an active exploratory intelligence emerging from a common evolutionary history.

In the 1960s Ilya Prigogine developed the theory of dissipative structures, for which he won the Nobel Prize in Chemistry in 1977. The theory maintained that long before a state of equilibrium is reached in irreversible processes, orderly and stable systems can arise from more disordered systems via self-organization. The idea of self-organization, for example, one's self-organization of knowledge as Piaget observed, is central to the role we now ascribe to the causal learning cycle. A dissipative system, examples of which are a learner, a team, or cultural groups constructing knowledge and becoming more intelligent, is open to a larger environment and operates far from a static equilibrium, in essence, to mete out or *dissipate* captured and stored energy from a raw form to transformed forms. Prigogine's ideas are discussed further in Chap. 6.

In individual learning, the transformed forms include memories, mental models, performance actions, knowledge, skills, and capabilities. The causal learning cycle utilizes dissipative processes to build intelligence and is proposed as the motive force behind not only individual learning in the traditional sense, but also, structural growth, skill development, memory, and other aspects of intelligence in social groups and culture at large.

A key issue at this juncture requires a distinction between energy and information, a distinction that Piaget did not directly address. The physical universe of which we are a part is fundamentally energy, as was made clear by Einstein (Einstein & Seelig, 1995). Distinct from physics, the universe of communications, meaning, and symbols is information, as made clear by Shannon and others (Shannon & Weaver, 1998). Both energy and information should be considered together as part of one evolving system when theorizing about the ontogeny of intelligence, cognition and learning processes. The discussion will deepen along these lines in the chapters that follow.

The problem is, how are energy and information connected and then also related to learning processes? In physics, the conservation of energy is a fundamental principle stating that energy cannot be created or destroyed; it can only change forms or be transferred from one system to another. This conservation principle ensures that the total energy in a closed system remains constant over time, even if it transforms between different types (e.g., kinetic, potential, thermal energy). Key phases of the causal learning cycle will make use of this principle, even though an individual human is an open system (e.g. takes in food and other inputs from the environment, transforms those inputs, and releases products into the environment). At times, the mind can operate like a closed system using previously stored stockpiles of energy (e.g. burning fats and sugars to power brain activity) as well as information stores (e.g. memories and cognitive strategies).

In contrast to energy, information theory (Shannon & Weaver, 1998) establishes that information is non-conserved. While it can be transmitted, processed, and transformed, information can also be created (e.g. multiplied nearly infinitely at a low cost of energy) or lost entirely, as when a scientific paradigm shift debunks a previous theory. Information losses or alterations might be due to factors like noise, errors, or entropy. This leaves the current project with a challenge to integrate both conserved and non-conserved resources into the theory of the causal learning cycle.

The connection between energy and information, while challenging is relevant in discussions involving both thermodynamics, such as turning food into bodies, bodily processes, and minds, and in discussions of information, such as turning signs from perception into semantics and meaningful symbolic motifs. In some contexts, there are analogies drawn between the conservation of energy in physics and certain principles in information theory, like the conservation of "bit" information or the relationship between entropy and information loss. These connections often highlight how the processes, transmission, and transformation of information involve energy expenditures and how information changes can be related to changes in the thermodynamic properties of a system, leading to discussions about the thermodynamics of computation and information processing. We discuss these aspects in detail in Chap. 6.

To continue, we now provide initial indications of the inclusion of energy and information, irreversibility, dissipation, and time into a conceptual foundation for the causal learning cycle by elaborating on Piaget's insights. Reviewing Piaget's four phases of cognition, we find:

4.2.1 Disequilibration 'Notices'

Disequilibration is a state of cognitive imbalance or discomfort experienced by individuals when they encounter new information or experiences from the external world that do not align with their existing mental schemas or understanding of the world. Disequilibration sets in motion a need to fill a gap between an existing mental model and an unidentified disturbance that is incoming via perception. In new terms, it is a *high-energy, low-information transformation* process that expands the learner's perceptual and psychological space. An example is hearing a sudden unidentified noise. Most of the nerves involved in hearing are carrying signals from the brain outward to the periphery, so most of the energy is maintaining an aural background, against which any disturbance is a small but noticeable difference in expectation. Low information now reflects the semantic value of the first-noticed sound, which has not yet been identified. Note that both energy and information are potentially lining up to be transformed by being 'noticed' against the background that is maintaining the world as we know it, in physical and symbolic terms.

4.2.2 Assimilation 'Remembers and Interprets'

Assimilation is the process of incorporating new information or experiences into existing mental frameworks, memories, or schemas without altering those frameworks significantly. The new sounds might be heard every day at about this time; memory matches the sound with a familiar object from experience. Assimilation by individuals thus *interprets* new information based on current understanding and is primarily an internal process of information retrieval or remembering. The assimilation phase is a *low-energy, high-information* process that further expands the mental model space because most of the work is internal using working or episodic memory. This function of assimilation suggests that this phase often requires consciousness. According to (Budson et al., 2022) episodic memory may have arisen first in evolution, and then consciousness arose within it, to creatively combine and rearrange memories of prior events. Unconscious assimilation processes can also occur when individuals are exposed to new concepts or experiences that align closely with their pre-existing beliefs or mental frameworks. The unconscious process can happen subtly and gradually, shaping individuals' perspectives and understanding without them consciously recognizing any accommodation or integration of the new information. If the disturbance is understandable and can fit in with current notions (or is deemed inconsequential), the causal cycle may skip accommodation and go directly to equilibration. If not, then the cognitive system must move to accommodate the new information.

4.2.3 Accommodation 'Creates Adaptations'

Accommodation adapts existing mental frameworks or schemas to incorporate new information or experiences that cannot be assimilated. Accommodation involves altering one's cognitive structures to better fit the new information, an act that requires energetic constant comparison with the external world. Accommodation creates structure through *symbolization*, as Suzanne Langer noted in *Philosophy in a New Key* (Langer, 1942) that developed the idea of the 'virtual illusion' of art (e.g. music) being the direct non-verbal experience of qualia (e.g. time in music). The perhaps uniquely human act of *symbolization* turns experiences into meaning by compressing experience into symbols. More generally, as explained further in Chap. 6, the effort of creating new meaning is a *high-energy, low-information* process that compresses regularities of both external and internal signals into compact symbols and descriptions (Baum, 2003). Unconscious accommodation may occur when individuals encounter new experiences that contradict their existing beliefs or mental schemas, in which case their cognitive structures might subtly adjust to accommodate or make sense of these new experiences without explicit conscious recognition of the accommodation process. The proposed framework associates the action of creating new semantics and symbols as part of the core mechanism of accommodation. In the next chapter, we provide key ideas for the dynamic transitions between the Piagetian phases, then in Chap. 6, we define a theory for dynamic equations that causally propel the transitions and states.

4.2.4 Equilibration 'Satisfies and Maintains'

Equilibration returns the system to a momentary, dynamic (not static) state of active balance between assimilation and accommodation. Equilibration is a state with few if any cognitive conflicts. This phase is a *low energy, high information* primarily internal process, which further compresses experience into semantics by finding regularities in the existing symbol system and mental models. While individuals might consciously engage in problem-solving to achieve psychological equilibrium, there is also an innate aspect of equilibration that operates unconsciously, where the mind seeks to balance and reconcile new information with existing cognitive structures without deliberate awareness of the process. Such is the case, for example, in Troxler's fading (Martinez-Conde et al., 2004), which involves the perceptual disappearance of stimuli in the periphery while fixating. Or staring at a centrally located stimulus. The visual system needs and responds to change; constant small movements called saccades are required to keep stimulating the nerve cells responsible for vision.

The *dynamic* (e.g. background creating) nature of equilibration is illustrated by the phenomenon of sensory habituation, when sensory receptors in our body, such as those responsible for touch or pressure when sitting down, become less responsive or "tuned out" over time when they are continuously stimulated. This reduced sensitivity happens because the receptors adapt to constant or repetitive stimuli by decreasing their response *to maintain a focus on changes in the environment* rather than constant sensations. For instance, when sitting down, the initial sensation of pressure or touch against a surface might diminish or fade away as the nerves or sensory receptors adapt to the consistent stimulus of sitting. It's not so much that the nerves get "tired" but rather that they become less responsive to a constant sensation due to the *brain's filtering of unchanging stimuli to prioritize detecting new or changing sensory information*. We suspect that, as indicated by Piaget, equilibration is an active state of intelligent readiness involving active cognitive (both conscious and unconscious) operations.

The phases of Piagetian equilibration theory represent a 4-cycle dynamic process through which individuals, teams, organizations, and cultures adapt and refine their cognitive structures to accommodate new experiences and foster growth and development (Fig. 4.1). Each state is momentary or fleeting and passes directly to a transition to reach the next state, which can take time to process. The states may be 'structural processes' in the sense that they are the lingering or episodic memory of the network of influences that have settled into the state from the last causal cycle. Such a mechanism could be operating at multiple levels within each cognitive agent, as envisioned in the *Society of Mind* (Minsky, 1988) and *A Thousand Brains* (Hawkins, 2021).

Since we are interested in a framework for developing AI to assist with learning processes, we can illustrate the dynamic process of the learning cycle by discussing the causal loop around the four phases starting from *disequilibration* (Fig. 4.1). Later, we'll focus on how AI-enhanced learning analytics can assist each of these processes.

Envision a person interacting with a human-computer interface. An eye tracker records how long the person views an object, which can support an inference that the person notices something of interest. Trace data of those kinds of actions, as well as what the person says, makes or creates, can be captured by information systems and, together with learning analytical techniques, can be used to extract patterns of an individual's learning behavior such as the recognition of concepts or

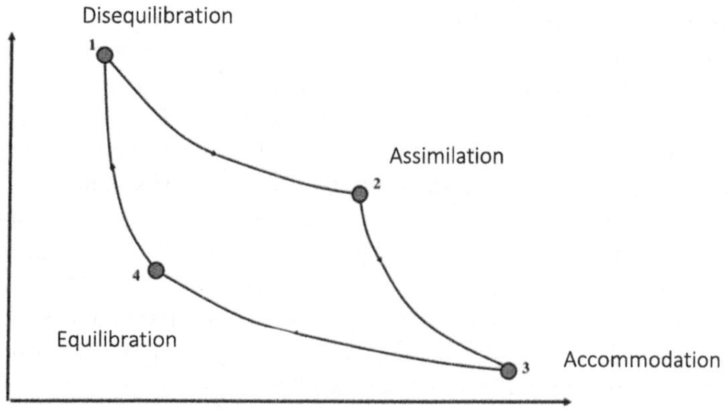

Fig. 4.1 Equilibration theory as a causal cycle. Note: The phases of the Piagetian learning cycle follow one another in a sequence that may begin at any node

giving responses that depend on prior knowledge; these may provide evidence of *assimilation*. Those patterns can be used to assess theory-informed concepts of interest in presenting and adapting a learning design (Ifenthaler et al., 2017) and together with real-time input from the person, can be further used to support the development of a personalized recommendation that offers an *accommodation* which AI can adroitly align with a learner's prerequisite knowledge and current performance to facilitate effective new learning. In this way, the assessment of individual outcomes can be enhanced by AI technologies that enable near-real-time interventions. Finally, evidence of the satisfaction of inquiries, questions, and practice attempts can trigger appropriate messages to the learner that the concept or skill has been grasped or correctly applied and returned to *equilibrium*. We review this cycle in depth in Chap. 7.

4.3 Limitations and Conclusions

This chapter did not extensively address the distinction between conserved energy and non-conserved information in understanding how intelligence arises as implied by the integration of Piaget and Kauffman's theories. Energy and information are connected to learning processes in that the creation and transmission of learning processes requires both, as seen in the brain's use of energy to send informational signals. The chapter focused on only one aspect of Piaget's theory; it did not discuss the developmental aspect of his theory, which has been covered in other psychological literature.

The chapter focused on how intelligence arises via a 4-phase cycle explained by Piaget's theory of equilibration and began to elaborate and explain a proposed series of causal learning processes. The chapter indicated how equilibration theory can be expanded to include new physical interpretations and dynamic transitions driving the causal learning cycle. The chapter projected that the cycle, in addition to being found in individual learning, will be found at the meso level in teams and organizations and cultural shifts at the macro level.

4.4 Moving Forward

The next chapter outlines four transition processes in the learning cycle by discussing the work of John Keller, who proposed the ARCS theory of motivation. By the end of the chapters in this section of the book, the narrative will have defined learning processes at the micro level—individual learning—by building on the foundation of four concepts in Piaget's theory of equilibration (Piaget, 1985) dynamically linked with transitions suggested by Keller's model (Keller, 2009) and unified with and interpreted broadly with Kauffman's theory of autocatalytic agency (Kauffman, 2000).

References

Baum, E. (2003). *What is thought*. MIT.

Bruner, J. (1997). Celebrating divergence: Piaget and Vygotsky. *Human Development*. https://doi.org/10.1159/000278705

Budson, A. E., Richman, K. A., & Kensinger, E. A. (2022). Consciousness as a memory system. *Cognitive and Behavioral Neurology, 35*(4), 263–297. https://doi.org/10.1097/WNN.0000000000000319

Drossel, B. (2023). The passage of time and top-down causation. In *The sixteenth Marcel Grossmann meeting* (pp. 3631–3645). https://doi.org/10.1142/9789811269776_0300

Einstein, A., & Seelig, C. (1995). *Ideas and opinions* (3rd ed.). Crown Trade Paperbacks.

Hawkins, J. (2021). *A thousand brains: A new theory of intelligence*. Basic Books.

Ifenthaler, D., Gibson, D., & Dobozy, E. (2017). The synergistic and dynamic relationship between learning design and learning analytics. *ASCILITE, 2017*, 1–5.

Kauffman, S. (2000). *Investigations*. Oxford University Press.

Keller, J. M. (2009). *Motivational design for learning and performance: The ARCS model approach* (1st ed.). Springer.

Langer, S. K. (1942). *Philosophy in a new key; a study in the symbolism of reason rite and art*. Harvard University Press.

Martinez-Conde, S., Macknik, S. L., & Hubel, D. H. (2004). The role of fixational eye movements in visual perception. *Nature Reviews Neuroscience, 5*(3), 229–240. https://doi.org/10.1038/nrn1348

Minsky, M. (1988). *The society of mind* (6. Pb-pr). Simon and Schuster.

Petraccone, L. (2023). Planetary entropy production as a thermodynamic constraint for exoplanet habitability. *Monthly Notices of the Royal Astronomical Society, 527*(3), 5547–5552. https://doi.org/10.1093/mnras/stad3526

Piaget, J. (1946). *The Child's conception of time* (0th ed.). Routledge. https://doi.org/10.4324/9780203715642

Piaget, J. (1985). *The equilibration of cognitive structures: The central problem of intellectual development*. University of Chicago Press.

Putnam, H. (1992). *Renewing philosophy*. Harvard University Press.

Shannon, C. E., & Weaver, W. (1998). *The mathematical theory of communication*. University of Illinois Press.

Chapter 5
Keller and the ARCS Model of Motivation

5.1 Introduction

ARCS Theory, standing for Attention, Relevance, Confidence, and Satisfaction, emerged in the late 1970s as a response to the growing need for a comprehensive framework to understand and enhance learner motivation in educational settings. Developed by John Keller, a prominent educational psychologist, the theory was a product of extensive research in the field of instructional design. Keller recognized the significance of motivation in learning outcomes and aimed to create a structured model that educators could utilize to design more engaging and effective learning experiences. The theory evolved through Keller's dedication to understanding how motivation functions in learning and teaching.

ARCS Theory comprises four fundamental components that form the backbone of motivational strategies in learning processes. **Attention** focuses on capturing learners' interest and concentration, utilizing methods such as stimulating curiosity or employing multimedia to engage learners. **Relevance** emphasizes connecting learning content to learners' experiences, interests, and future goals, making the material meaningful and applicable. **Confidence** involves building learners' self-assurance and belief in their abilities, fostering through construction, a sense of competence and autonomy. Lastly, **Satisfaction** revolves around reinforcing and consolidating learning through feedback, rewards, or a sense of accomplishment, ensuring learners feel fulfilled and motivated to continue their educational journey. In our current project to re-think learning theory foundations, these four action components (e.g. capturing, connecting, constructing, consolidating) are associated with the causal links in the network model of the learning cycle while the nodes remain associated with Piaget's 4-phase theory of equilibration (Fig. 5.1).

Keller emphasized several key points regarding the development and significance of the theory. He believed that designing instructional materials and activities solely based on content wasn't sufficient; motivation should be at the center of

© The Author(s), under exclusive license to Springer Nature Switzerland AG 2024

D. C. Gibson, D. Ifenthaler, *Computational Learning Theories*, Advances in Analytics for Learning and Teaching,
https://doi.org/10.1007/978-3-031-65898-3_5

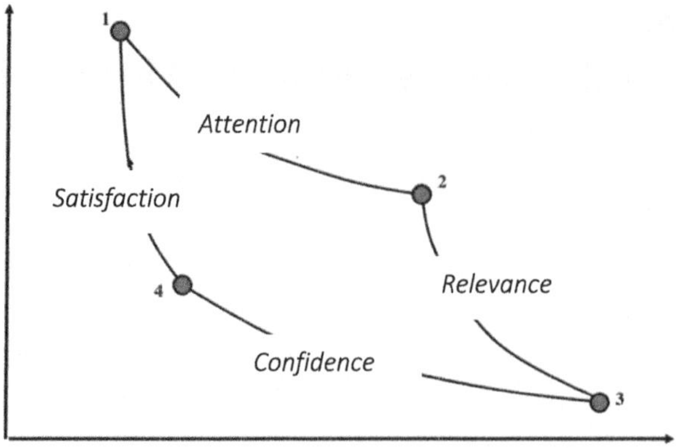

Fig. 5.1 Keller's ARCS Theory components as causal links in the learning cycle

instructional design. The ARCS model thus aimed to provide a framework that incorporated motivational strategies into the design process as a systematic approach to address the four essential components of motivation. Keller emphasized that the model wasn't a rigid formula but rather a flexible framework that could be adapted across various learning contexts and subjects. He encouraged educators and instructional designers to apply the ARCS principles creatively to cater to diverse learner needs and preferences. His goal was to improve learning outcomes by engaging students more deeply with content, leading to better retention, understanding, and application of knowledge.

In our current project, the four components of the ARCS theory are offered as explanatory propositions characterizing the causal linkages among the Piagetian ideas arranged as a causal chain. The next sections attempt to explain.

5.2 Models of Causation

In scientific writing, causal models are employed to systematically represent and analyze relationships, which in the current project are represented by the links between nodes as well as the general topology of the entire micro-level network (e.g. the relationship of nodes to each other in both abstract space and temporal order). This chapter proposes that Keller's motivational components' can be used as broad labels for the driving forces or causes that propel the transitions in the cyclical Piagetian nodes of the central learning mechanism. Causation, in this most fundamental sense, refers to the relationships that drive one event (the cause) to bring about another event (the effect). Understanding causation is essential for explaining and predicting outcomes in complex systems but modeling the concept needs to be understood from both linear and nonlinear points of view, particularly in the case of cyclic attractors driving the dynamics of the system, as asserted for the learning cycle. The rest of this section will briefly define and attempt to clarify what is meant here by causation.

The process of proving or establishing causal models involves a combination of empirical evidence, statistical analyses, and adherence to rigorous scientific standards. Common approaches to validate and support causal models include observational studies and experiments in which researchers observe and analyze real-world situations to identify patterns and associations, and controlled experiments that involve manipulating variables and observing the effects. Randomized controlled trials, for example, are particularly powerful in establishing causation by minimizing confounding variables and allowing for causal inference. Additional approaches to validating a causal model may include analysis of the correlation and covariation between the presumed cause and effect as an initial indicator of a potential causal relationship. Crucially, temporal order is key in demonstrating causation. The cause must precede the effect in time, except in the special case of Granger causality (Amblard & Michel, 2011; Seth, 2007) as found in a complex periodic or quasi-periodic system, where cyclic patterns of pre and post-events occur on a semiregular and recurring basis.

One prevalent and orthodox type of causality is the deterministic model, which asserts that causes *uniquely* determine their effects. This model operates under the assumption that given the same initial conditions, the same outcome will always follow. However, the deterministic model may oversimplify reality by neglecting stochastic elements and uncertainties present in many systems. In cyclical systems, for example, temporal order and nonlinearities may interfere with some of the requirements of determinism such as prediction accuracy and the universality of the underlying dynamics (e.g. universality states that the rules won't change over time, but in complex systems change is ever-present). In the current project, the causal learning cycle model makes use of a weak form of determinism in which a cause *substantially* determines an effect (rather than *uniquely*) and the context of the system (e.g. its current state of evolution with embodied potentials) may render the causative force impotent or nonexistent in certain circumstances.

Another widely used modeling approach is founded on probabilistic or statistical assumptions, which acknowledge the inherent uncertainties in real-world situations. Probabilistic models assert that a cause *increases the likelihood of an effect* but do not guarantee it. While this approach is more realistic in capturing the complexity of real-world scenarios, it introduces challenges in accurately estimating and interpreting probabilities. Choosing 'priors' in Bayesian analysis (Jaynes, 1976), for example, is evidence of the challenge of estimation, and concerns about 'explainable AI' decisions in machine learning (Khosravi et al., 2022) is an example of the challenge of interpretability.

Granger causality (Marinazzo et al., 2011), a special case within statistical models, focuses on time-series data in quasi-cyclical systems and assesses the extent to which past values of one variable can predict future values of another. It identifies predictive patterns in complex systems (Sugihara et al., 2012) and is thus highly relevant to the learning cycle model proposed here.

Each of these causal modeling approaches has its strengths and limitations. Deterministic models provide clarity and simplicity but may oversimplify dynamic systems. Probabilistic models offer a more nuanced understanding but may not fully capture or explain the mechanics or details of intricate dependencies. Granger causality, while useful for time-series analysis and providing highly accurate transition rules for dynamic systems, should be applied cautiously in evolving systems as causes may drift as the context of an effect may change.

Each modeling approach has limitations in fully establishing causation. Each requires additional external review, counterfactual arguments, analysis of consistency with theory, meta-analyses with extant literature, expert review and other kinds of validation. Overall, the choice of a causal model depends on the specific context and goals of the analysis, with careful consideration of the assumptions and constraints inherent to each model type.

With these issues in mind, the Keller 'motivational components' might best be thought of as labels for causative forces that lie on each specific link in the Piagetian network representation of the learning cycle, adding a new layer of information to help explain the transitions in the model. In the next chapter, physical mechanisms from computational biology are added to both Piaget and Keller to give a more detailed view of the ecosystem of causes in the complete micro-level model.

5.3 Motivation and the Causal Learning Cycle

Because Keller's approach was more focused on practical application than explicit theoretical citations, his works often emphasized practical strategies for engaging learners' attention, drawing from empirical evidence and instructional design principles rather than explicitly stating reliance on specific theoretical frameworks and empirical findings. Nonetheless, the ARCS model is consistent with several theories and empirical evidence related to learning processes that will be outlined below. The cognitive leap from practicality to theoretical significance emphasizes that the tenets of motivation observable in individual learning are representations of causative forces propelling the learning cycle from state to state. In a sense, each ARCS component motivates one state to transition into the next state and thus each partially explains the local causative force on its link between the Piagetian states.

Attention, for example, has support from theories of Arousal, Dual Coding, Cognitive Load, and the Cognitive Theory of Multimedia Learning. Arousal Theory, particularly the Yerkes-Dodson Law (Teigen, 1994), suggests that moderate levels of arousal or attention are optimal for performance. However, as the input channels of perception can become overloaded when too much information is presented (as well as unresponsive when arousal levels are too low), theorists have proposed models for divided and selective attention. Alan Paivio's dual coding theory (Paivio, 1991) and John Sweller's cognitive load theory (Sweller et al., 2011) are both foundational for example, in Richard Mayer's cognitive theory of multimedia learning (Mayer, 2009), which is based on empirical findings that learners can better understand an explanation when it is presented in words and pictures than with words alone. The current project extends the idea of managing a finite bandwidth of attention and processing to additional multisensory channels like tactile, haptic, emotional and metacognitive processes, utilizing both signals and semantics involved in bringing items to attention (Azevedo & Gašević, 2019).

Attention can be defined computationally in a network model of intelligence arising in conjunction with distance or similarity measures that compare extant and incoming patterns. Also of interest is the development of transformer models, key to large language models in natural language processing, which introduce a computational role for attention, declaring that is 'all you need' (Vaswani et al., 2017) for signs of intelligence (e.g. dispensing with recurrence and convolutional networks). In the current project, attention mechanisms are offered as explanatory theories of the causal conditions that drive Disequilibration to morph or transition into Assimilation.

Relevance overlaps somewhat with selective attention mechanisms and is supported by psychological theories like Broadbent's Filter Model (Broadbent, 1958) and Treisman's Attenuation Theory (Treisman, 1964), which underscore the selective nature of attention, proposing that individuals filter and prioritize information based on relevance. Mechanisms at the neural network level for creating relevance include the quasi-homomorphism proposed by John Holland for complex adaptive systems (Holland, 2019). In a quasi-homomorphism network structure (e.g. an inexact or incomplete mapping of sensory inputs as a representation), parallel rules about the world, such as the configurations of simultaneously firing sensory and cognitive elements in a mental model, are always 'on' unless an exceptional circumstance requires a carve-out for an identifiable subset of signals in the network. The exception rule is then, in that specific context, locally relevant to the current context along with the exception signals. The rule or a decision based on it only comes into play when the right combination of signals arises while the larger network is simultaneously globally relevant. In this way, the relevancy of context shapes the semantics of a mental model or neural net activation assembly or ensemble.

A network theory mechanism for this interpretation of relevancy may be the *superposition of polysemantic nodes* (Elhage et al., 2022) and *superposition of models* (Cheung et al., 2019), theories that support multiple encodings needed for a neural net vision of relevance. When network features are sparsely represented by nodes, superposition allows compression beyond what a linear model can do, at the cost of interference that requires nonlinear filtering by the quasi-homomorphism. In a multi-nodal representation needed for a more complex model, it is possible to partially exploit the excess capacity present in the neural network during training by *learning multiple tasks* (with forward implications for expertise and confidence to be discussed next). Hyperdimensional representations (Kanerva, 2009) can co-exist in the superposition of models and can be keyed to apply when appropriate conditions are present for classification or other activations. Thus, relevance can be defined computationally in network models of intelligence. In our current project, these mechanisms are offered as explanatory theories of the motivic forces that cause Assimilation to transform into Accommodation.

Confidence aligns with Albert Bandura's Self-Efficacy Theory (Bandura, 1997). The notion that learners need to feel capable and confident in their abilities relates to Bandura's emphasis on self-belief and perceived competence. It also aligns with theories of expertise (Ericsson et al., 2006; Sternberg, 1999) that indicate that practice, when coupled with an authentic situation calling for the expertise-in-use and accompanied by effective coaching, appears to predict the rate and ultimately the limits of the development of expertise within a given domain. In our current project to establish a computational biology model for multiscale learning, at the micro level, we can re-interpret *practice* as the reinforcing of Hebbian synapses, *authentic situations* as the adjacent resonating network of Bayesian weights of nearby pre- and post-synaptic junctions, and *coaching* as contextual feedback from the rest of the 3-level hierarchical network estimating the fitness of the response, for example as homologous functions of socially observed tenets of confidence. Expertise and confidence can be computationally defined along the lines of evidence-centered design (Mislevy et al., 1999) in which complex profiles of behavior (and with AI-enhanced multisensory interfaces, behaviors include emotion, physical signals, metacognition, and other cognitive processes) and rules of evidence are used to support inferences about what someone knows and can do. In the current project, the signal-to-semantic evidential structure is offered as an explanatory theory of the motivation that causes Accommodation to transform into Equilibration.

Satisfaction through the provision of feedback and reinforcement is supported by Achievement Motivation Theory (McClelland & Burnham, 1980), which highlights the role of feedback, recognition, and reinforcement in fulfilling an individual's need for achievement and linking success to internal satisfaction. In addition, the concept of extrinsic motivation in Self-Determination Theory (Ryan & Deci, 2000) suggests that external rewards and recognition can foster satisfaction and can either support or undermine individuals' intrinsic motivation and overall well-being, while intrinsic motivation underscores the importance of a sense of achievement and enjoyment in learning. In the current project, the idea of the return to equilibration (with slight updates due to recent experience) involves resetting the weights of the pre- and post-synaptic junctions of each node and link in a mental model as well as in each multi-task composition of superimposed network representations, or model. The updating of weights resets the Equilibration of the network, presumably to the lowest maintenance energy level feasible to sustain the information flows (Bogacz, 2017), to serve as the next reference framework for comparison with newly arising inputs.

5.4 Limitations and Conclusions

This chapter did not discuss nor compare theories of motivation other than Keller's ARCS theory. It did not, for example, delve into social theories of motivation and their emphasis on social interactions or the influence of social factors on culture, which will appear in later chapters. The chapter limited itself to characterizing the linkages in the micro level's causal learning cycle network representation, rather than providing a comprehensive comparison of motivation theories.

In conclusion, the ARCS model, originally proposed by Keller as a flexible framework for designing motivational instructional materials, contains four concepts that the current project uses to characterize the causative forces driving the transitions from disequilibration to assimilation to accommodation to re-equilibration. Attention mechanisms triggered by disequilibration drive toward assimilation. Relevance mechanisms drive toward accommodation. Confidence and expertise drive toward re-equilibration. A physical theory of the dynamics will be provided in the next chapter.

5.5 Moving Forward

The goal of this chapter was to characterize the linkages in the network model of the causal learning cycle with labels from the ARCS theory of motivation. Later, in Chaps. 10, 11, and 13 the narrative will include social theories and their emphasis on social interactions, learning from others, and the influence of social factors on culture. At these later points, the homologous causative functions motivating state transitions in the learning cycle will re-appear as social and then cultural processes.

For example, Social Cognitive Theory, developed by Albert Bandura (Bandura, 1986), highlights the role of observational learning, self-efficacy, and the influence of social environments in shaping behavior and motivation. It suggests that individuals learn by observing others and that their motivation and actions are influenced by the modeling of behaviors, attitudes, and outcomes demonstrated by people around them.

In the next chapter, the learning cycle model is extended by computational biology concepts that support the emergence of intelligence within a network model of the expansion of agency into the *adjacent possible*, as developed by Stuart Kauffman, supported by Ilya Prigogine's theory of dissipative *structure*.

References

Amblard, P.-O., & Michel, O. J. J. (2011). On directed information theory and Granger causality graphs. *Journal of Computational Neuroscience, 30*, 7–16. https://doi.org/10.1007/s10827-010-0231-x

Azevedo, R., & Gašević, D. (2019). Analyzing multimodal multichannel data about self-regulated learning with advanced learning technologies: Issues and challenges. *Computers in Human Behavior, 96*, 207–210. https://doi.org/10.1016/j.chb.2019.03.025

Bandura, A. (1986). *Social foundations of thought and action: A social-cognitive theory*. Prentice-Hall.

Bandura, A. (1997). *Self-efficacy: The exercise of control*. W. H. Freeman.

Bogacz, R. (2017). A tutorial on the free-energy framework for modelling perception and learning. *Journal of Mathematical Psychology, 76*, 198–211. https://doi.org/10.1016/j.jmp.2015.11.003

Broadbent, D. E. (1958). *Perception and communication*. Pergamon Press.

Cheung, B., Terekhov, A., Chen, Y., Agrawal, P., & Olshausen, B. (2019). Superposition of many models into one. *ArXiv*. http://arxiv.org/abs/1902.05522

Elhage, N., Hume, T., Olsson, C., & Schiefer, N. (2022). *Toy models of superposition*. Transformer Circuits Thread.

Ericsson, K. A., Charness, N., Hoffman, R. R., & Feltovich, P. (2006). *The Cambridge handbook on expertise and expert performance*. Cambridge University Press.

Holland, J. (2019). Complex adaptive systems: A primer. In D. Krakauer (Ed.), *Worlds hidden in plain sight: Thirty years of complexity thinking at the Santa Fe Institute*. Santa Fe Institute.

Jaynes, E. (1976). Confidence intervals vs. Bayesian intervals. In *foundations of probability theory, statistical inference, and statistical theories of science: Vol. II* (Issue 2, pp. 175–257). http://scholar.google.com/scholar?hl=en&btnG=Search&q=intitle:Confidence+intervals+vs+bayesian+intervals#0

Kanerva, P. (2009). Hyperdimensional computing: An introduction to computing in distributed representation with high-dimensional random vectors. *Cognitive Computation, 1*(2), 139–159. https://doi.org/10.1007/s12559-009-9009-8

Khosravi, H., Shum, S. B., Chen, G., Conati, C., Gasevic, D., Kay, J., Knight, S., Martinez-Maldonado, R., Sadiq, S., & Tsai, Y.-S. (2022). Explainable artificial intelligence in education. *Computers and Education: Artificial Intelligence*, 100074. https://doi.org/10.1016/J.CAEAI.2022.100074

Marinazzo, D., Liao, W., Chen, H., & Stramaglia, S. (2011). Nonlinear connectivity by Granger causality. *NeuroImage, 58*, 330–338. https://doi.org/10.1016/j.neuroimage.2010.01.099

Mayer, R. E. (2009). *Multimedia learning* (2nd ed.). Cambridge University Press.

McClelland, D. C., & Burnham, D. H. (1980). Power is the great motivator. In H. J. Levitt, L. R. Pondy, & D. M. Boje (Eds.), *Readings in managerial psychology* (3rd ed.). University of Chicago Press.

Mislevy, R., Steinberg, L., & Almond, R. (1999). *Evidence-centered assessment design*. http://www.education.umd.edu/EDMS/mislevy/papers/ECD_overview.html

Paivio, A. (1991). Dual coding theory: Retrospect and current status. *Canadian Journal of Psychology/Revue Canadienne de Psychologie, 45*(3), 255–287. https://doi.org/10.1037/h0084295

Ryan, R., & Deci, E. (2000). Self-determination theory and the facilitation of intrinsic motivation, social development, and well-being. *Contemporary Educational Psycology, 25*, 54–67.

Seth, A. (2007). Granger causality. *Scholarpedia, 2*, 1667. https://doi.org/10.4249/scholarpedia.1667

Sternberg, R. J. (1999). Intelligence as developing expertise. *Contemporary Educational Psychology, 24*(4). https://doi.org/10.1006/ceps.1998.0998

Sugihara, G., May, R., Ye, H., Hsieh, C., Deyle, E., Fogarty, M., & Munch, S. (2012). Detecting causality in complex ecosystems. *Science New York, N.Y., 338*(6106), 496–500. https://doi.org/10.1126/science.1227079

Sweller, J., Ayres, P., & Kalyuga, S. (2011). *Cognitive load theory*. Springer. https://doi.org/10.1007/978-1-4419-8126-4

Teigen, K. H. (1994). Yerkes-Dodson: A law for all seasons. *Theory & Psychology, 4*(4), 525–547. https://doi.org/10.1177/0959354394044004

Treisman, A. (1964). Monitoring and storage of irrelevant messages in selective attention. *Journal of Verbal Learning and Verbal Behavior, 3*(6), 449–459. https://doi.org/10.1016/S0022-5371(64)80015-3

Vaswani, A., Shazeer, N., Parmar, N., Uszkoreit, J., Jones, L., Gomez, A. N., Kaiser, L., & Polosukhin, I. (2017). *Attention is all you need*. https://doi.org/10.48550/ARXIV.1706.03762

Chapter 6
Complexity Theory and Learning

6.1 Introduction

This chapter introduces Ilya Prigogine's theory of *dissipative structures* (Prigogine, 1976) and Stuart Kauffman's theory of *autocatalysis* (Kauffman, 2000) and links those ideas to a complexity theory of learning by adding the concepts to the learning cycle model. The chapter will then indicate how the completed causal learning cycle model of learning forms a foundational mechanism for learning at micro, meso, and macro levels. In Chaps. 4 and 5 we discussed Piaget's and Keller's theories of intelligence and behavior and brought the theories together into a network model with Piagetian nodes and Keller edges in a cycle. Each node and its out-transition leads to and influences a nearby causally connected node in the network; its incoming transition comes from a previous cause in the cycle. This helps explain the internal structure of individual learning processes to some extent, but toward what end? Why is there a natural drive to learn? How do agency and self-awareness emerge?

In this chapter, we add a physical mechanism that helps explain the ontogeny of intelligence with roots in both thermodynamics and biology. Thermodynamics and biology are involved to account for the learner's intake and use of food, information, and other resources to meet physical and psychological needs including building a brain and mind. In addition, the learner's intake of information is needed for constructing a social world of family and work teams, as well as for participating in a broader cultural environment. Resources from the environment as well as the internal world of the learner are stored as reservoirs of potential energy and information for *dissipative structures* (Prigogine, 1976) that include cognitive processes, such as nonverbal know-how, memories, mental models and expertise, as well as general knowledge of self, others and the world, and overall, the development of intelligence.

© The Author(s), under exclusive license to Springer Nature Switzerland AG 2024

D. C. Gibson, D. Ifenthaler, *Computational Learning Theories*, Advances in Analytics for Learning and Teaching,
https://doi.org/10.1007/978-3-031-65898-3_6

6.2 Ilya Prigogine

Ilya Prigogine was born in Moscow, Russia on January 25, 1917. He obtained both his undergraduate and graduate education in chemistry at the Universite Libre de Bruxelles. He moved to Texas in 1959 and became Regental Professor and Ashbel Smith Professor of Physics and Chemical Engineering at the University of Texas at Austin. Prigogine was awarded the Nobel Prize in chemistry in 1977 for his contributions to nonequilibrium thermodynamics, particularly the theory of dissipative structures.

The main theme of his scientific work has led to a better understanding of the role of time in the physical sciences and biology, illustrated by his book with Isabelle Stenger "Order Out of Chaos," first published in 1978. The book challenged traditional views of science by highlighting the nature of nonlinear dynamics and the behavior of systems far from equilibrium. Certain systems, when pushed away from equilibrium, can undergo spontaneous self-organization, leading to the emergence of novel and unexpected patterns, structures, and behaviors. Such systems that are open to energy and matter flows can create organized patterns or structures amid chaos—*dissipative structures*. His theory provides a mechanism for the creation of order (and thus life and intelligence according to Kauffman), within inherently noisy, dynamic, and evolving systems. Prigogine's life work significantly contributed to our understanding of how complexity, far-from-equilibrium conditions, and nonlinearity play integral roles in shaping the dynamics of natural systems, challenging the traditional Newtonian view of deterministic and predictable processes in nature, such as the second law of thermodynamics. The second law of thermodynamics, you may recall, states that in any isolated physical system, order inevitably dissolves into decay. But Prigogine showed that in an open system powered by an energy source, such as the Earth exposed to heat from the Sun or a cognitive system (e.g. a learning agent) exposed to sensory inputs, structures can evolve and become more complex.

Dissipative structure theory fundamentally altered the science of how systems evolve and organize and highlighted the potential for order and stability to arise in irreversible non-equilibrium conditions. This concept profoundly impacted scientific thinking by emphasizing the importance of non-linear dynamics and self-organization in understanding the behavior and evolution of complex systems. After introducing Stuart Kauffman's work, we'll summarize the key points of Prigogine's and draw applications from both to educational theories.

6.3 Stuart Kauffman

Stuart Kauffman, born in 1939, is a prominent figure in the fields of theoretical biology, complex systems, and the origins of life. He earned his medical degree from the University of California, San Francisco, and later pursued postdoctoral research in developmental biology at the University of Oxford. Kauffman's interdisciplinary journey led him to propose several aspects of complex adaptive systems that remain significant contributions to theoretical biology and complexity theory. He is renowned for his exploration of the complexity of biological systems, proposing the concept of *order for free* through self-organization in complex systems, emphasizing how order can emerge spontaneously without external direction. Kauffman's work on Boolean networks and the exploration of the fitness landscape profoundly influenced the understanding of biological evolution and the origins of life. His interdisciplinary approach and pioneering ideas have left an indelible mark on fields spanning biology, complex systems, and the philosophical understanding of emergence and self-organization.

Kauffman defined the cell (e.g. a neuron) as a complex network, a paradigm that revolutionized the understanding of cellular biology (Kauffman & McCullogh, 1967). He conceptualized cells not as individual entities but as intricate networks of interacting molecules, where the dynamic relationships and connections govern the cell types and behaviors. For example, he described *cell types as dynamical attractors* (recall this term from Chap. 1) in gene regulatory networks and *cell differentiation as transitions between attractors* (e.g. the edges of a network representation like the 3-tier model). Kauffman's theory emphasizes that cellular functions and behaviors arise not solely from the actions of individual components but from the collective interactions within a network. His network-centric approach underscores the significance of interconnectedness and emergent properties within cells, reshaping understanding of the fundamental nature of cellular life and its inherent complexity.

Stuart Kauffman's concept of *expanding into the adjacent possible* refers to the evolutionary process where systems, whether biological or otherwise, continuously explore and expand their possibilities or potential states by moving incrementally into adjacent, feasible configurations that are adjacent to their current state. Expansion into the adjacent possible is the mechanism through which evolution explores new opportunities, preadapts to changing environments, and generates novel possibilities. It allows biological systems to incrementally explore neighboring, reachable configurations or states from their current condition. This exploration into adjacent possibilities enables systems to diversify, innovate, and adapt by discovering new functionalities, structures, or behaviors that were previously unexplored.

Thus, the inherent drive to expand into the adjacent possible represents a causal force behind the creative imperative of evolution, as it facilitates the emergence of new traits, species, and biological complexities through continuous exploration and expansion into nearby, viable states. It underlines the dynamic nature of evolution, emphasizing the system's capacity to generate novelty, adapt to changing circumstances, and explore new solutions, driving the ongoing innovation and complexity observed in biological systems. In the current project, we propose that expansion into the adjacent possible during individual learning is responsible for not only the emergence of intelligence but also agency—the 'power to initiate action' (Bandura, 2001)—and self-awareness. The marriage of agential intention and self-awareness is central to self-regulated learning and metacognition (Dinsmore et al., 2008) and will be discussed in the introduction to Chap. 7 as goals for the development of AI applications that promote individual learning processes.

According to Kauffman, evolution is not merely a linear progression or a predetermined path but a process of continuous exploration and innovation within the realm of what is feasible or accessible from the current state. We will adopt this idea into the causal learning cycle and assert that the urge to learn is universal, as it is based on this basic principle of expansion into the adjacent possible (c.f. compare with what Vygotsky called the 'zone of proximal development') and that the expansion is driven by a causal cycle represented by the amalgam of Piaget-Keller-Kauffman (PKK) theory (Fig. 6.3).

To continue the story, we first review Kauffman's logic, based on an idealization of a thermodynamic engine (Kauffman, 2000). Kauffman's insight was that the relationships of thermodynamics, irreversibility, and information theory to entropy were bound up with the emergence of self-catalyzing chemical networks, and those, in turn, were foundations for the emergence of life and then intelligence from the primordial soup. He extended the idea of exploration and exploitation of niches to account for *pre-adaptations* (e.g. how nature seems to know ahead of time that an eye, for example, might be a useful sensor) and this concept became the basis for the expansion into the adjacent possible, which we take to be a useful model for the universal motivation for learning processes. While the aim of creative universal expansion in Kauffman is the emergence of life, intelligence, and the survival of species, the current project applies the creative expansion concept to the emergence of agency and self-awareness at the individual level, the emergence of collective knowledge and awareness at the social level, and the emergence of global multiculturalism at the macro level. These themes are taken up in subsequent chapters.

Stuart Kauffman drew upon Carnot's insights concerning thermodynamics, particularly the theory of an ideal heat engine, to describe autocatalytic cycles in biological systems. Kauffman applied this analogy to explain how autocatalytic cycles operate and sustain themselves. An ideal heat engine operates cyclically, transferring heat from a high-temperature reservoir to a low-temperature reservoir to perform work. The dissipation of energy during this transfer provides energy for life to work against the second law of thermodynamics (e.g. to fight against the slide into heat death as the universe is winding down) as an agent transfers potential energy from a storehouse into kinetic work needed to sustain life and intelligence.

In an autocatalytic cycle, according to Kauffman, certain molecules or chemical reactions drive a self-sustaining loop where the products of one step in a process catalyze subsequent reactions, which in turn produce more of the original catalysts. Just as the heat engine sustains work cycles, the autocatalytic cycle in biological systems sustains a sequence of reactions catalyzed by the products of previous steps, demonstrating a form of self-replication or self-perpetuation. This analogy helped Kauffman illustrate how certain chemical systems or molecular interactions exhibit cyclic, self-sustaining behavior akin to the operations of thermodynamic cycles.

To introduce the ideal thermodynamic engine as a metaphor for the mechanism propelling the learning cycle, take a breath, hold it, and release it slowly before letting out the last few puffs of air. Count to 4 while slowly breathing in; this, in Carnot's terms is the '*isothermal expansion*' phase in which new inputs expand and warm the space. Then quietly, not forcefully, hold your breath and count to 2; this is the '*adiabatic expansion*' phase in which no new outside energy is added but there is still work being done even as the temperature drops (n.b. the 'free energy' framework (Bogacz, 2017) for modeling perception and learning applies here). Now, slowly let out the air while counting to 4 again; this is the '*isothermal compression*' stage in which the release of energy temporarily stored from the in-breath can be used to do work, such as blow up a balloon or turn a pinwheel. Finally, release the last remaining air; this is the '*adiabatic compression*' moment or stage. The next breath is then automatic as you gulp to take in more air. The automation of this final return to the isothermal expansion phase is a hint of what is meant by 'autocatalysis' in that the previous moment creates the conditions for the environment to push back. And the cycle continues.

As a model of the learning cycle, we propose that each stage sets up the conditions for each following stage and inexorably leads to it, leading to the idea of the *causal learning cycle*. We discuss this further below by returning to the ideas of Piaget and Keller and adding Prigogine's and Kauffman's insights.

6.4 Physics, Information, and Autocatalysis

To frame the matter, we'll discuss three models: control systems, smart systems, and learning systems. In physical control systems, the intelligence of some process is embedded in natural physical phenomena determined by energy and the laws of thermodynamics; in smart systems, information theory is added to the model; and in learning systems, evolutionary and computational processes combine and are added to create and grow intelligence.

6.4.1 Physical Control Systems

A simple model that will illustrate a physical control mechanism is an old-fashioned room thermometer set to keep a room warm in winter. The room is partially insulated, but there are also windows and doors, so as people come and go, the temperature fluctuates, and the room gets colder if the heater is off. The old thermometers relied on a curled piece of metal wire, at the end of which, was a sealed tube with mercury and an incoming and outgoing wire connected to the heating system. A setpoint for turning on the heater functioned as a 'point attractor' of the system (Chap. 1) and was a physical location on a fixed dial of possible temperatures at which the tube would be tipped to allow the mercury to flow to fill the gap between the wires, closing the circuit and sending electricity to the heater to turn on. When the room is colder than the set point, the wire contracts, raising the tube and causing the mercury to complete the circuit, which turns on the heater. When the room heats up, the wire relaxes, the mercury drops away from the wires and a gap is created that turns off the heater. There is a delay in the change from contracted to relaxed, which causes a delayed reaction in the system.

If we graphed the temperature of the room over time, it would look somewhat like a sine wave, passing through the setpoint three times (Fig. 6.1) creating four states. The names and numbering of the states are arbitrary, except that whichever state is called #1 must be followed by the next state in the causal cycle; the sequence of the states of the cycle cannot be mixed up or reversed, as this is an irreversible time-dependent process with a real-world history.

State 1: room warming, heater on, temperature above the setpoint (and rising)
State 2: room cooling, heater off, temperature above the setpoint (and falling)
State 3: room cooling, heater off, temperature below the setpoint (and falling)
State 4: room warming, heater on, temperature below the setpoint (and rising)

Note that this simple system contains several causal chains: the contracting wire reacting to cold room temperature causes the mercury to flow via gravity to close the circuit, the closed circuit causes the engine to turn on, the engine sends heat into the room, the relaxed wire reacting to high room temperature causes the mercury to leave the gap in the wires, the open circuit causes the heater to turn off, the room grows cold again and the cycle continues.

Note that if a measure of the system took place during one of the four states, an evaluation of the system would have four different outcomes (educational assessment practitioners take note!), as an assessment would associate efficiency and cost with the energy expense and fitness for purpose in turning the heat on when needed or vice versa.

State 1 is costly and unfit for purpose
State 2 is low-cost and fit for purpose
State 3 is low-cost but unfit for purpose
State 4 is costly but fit for purpose

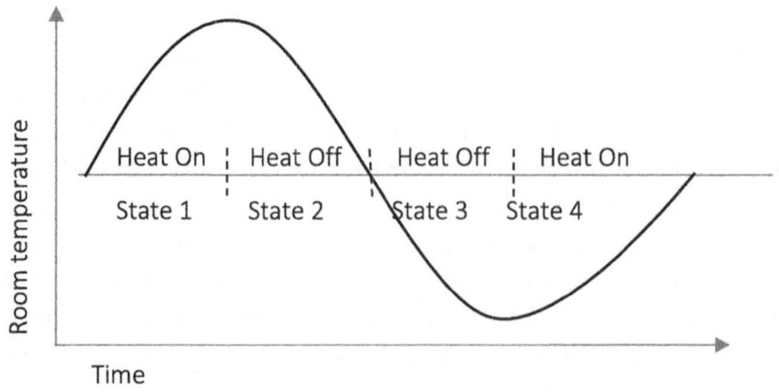

Fig. 6.1 A thermostat model of a mechanistic control system

It is perhaps no wonder that point-in-time measures of complex phenomena such as individual learning, team learning, and cultural change can result in conflicting and confusing research outcomes!

The room-thermostat-heater is an open system, enclosed in an encompassing environment of winter. No nonphysical (non-conserved) information is passing among the nodes (e.g. state 1, 2, 3, 4) nor during the transitions (e.g. the duration of each stage) of the system, instead, each node's geometry of behavior or topology (e.g. contracting, expanding, closing or opening the gap) changes the flow of energy from one form to another, as predicted by the law of the conservation of energy. The model of the causal learning cycle that we are developing will mirror many of these features, as there is also a flow of conserved energy taking place during learning. In the physical control model, there is no learning and no smart decisions based on information. Although we might say that the geometry of the physical components exhibits a form of embedded intelligence, there is no 'decision moment' based on logic and information.

6.4.2 Smart Systems

The second model—a smart system—adds and leverages information flows and logic into the control system network to shorten the delays, compress time, and lessen the expense of transforming physical energy into actions. Adding information flows and decision-making agency results in improved efficiency and fitness of the system. Some of the physical couplings of nodes and their transition processes can be replaced by information flows to logic circuits. Signals from one part of the process can now travel unimpeded and simultaneously to more than one receiving location because the non-conserved resource (information) can be shared without diminishment.

In the smart system model of the thermostat, we replace the mercury flowing down a gravity gradient, with a logical operation involving sensors and 'if…then' statements. For example, the logic unit might turn off the heater as soon as possible after the system enters State 1 and ensure that it stays off during State 2. The system can also turn the heater on as soon as possible after the system enters State 3 and ensure that it stays on during State 4. Although transmitting information costs some energy, it is a lot less expensive and time-consuming than the mechanical system, which has friction, thermodynamics, and the varying physical properties of the transmitting media. As a result of the addition of information-based decision-making, the system does not swing as far into states that are costly or unfit (Fig. 6.2).

Relating the smart model to the ontogeny of intelligence in the human brain, logic circuits are made possible by 'reentrant' neurological processes (Edelman & Mountcastle, 1982). That is, as the brain matures during development, additional neural connections are made, some of which result in loops re-entering the nerve cell body, axons, or dendrites as either pre- or post-synaptic connections. These self-referring loops can create the basic logical processes of *summation*, *and*, *or*,

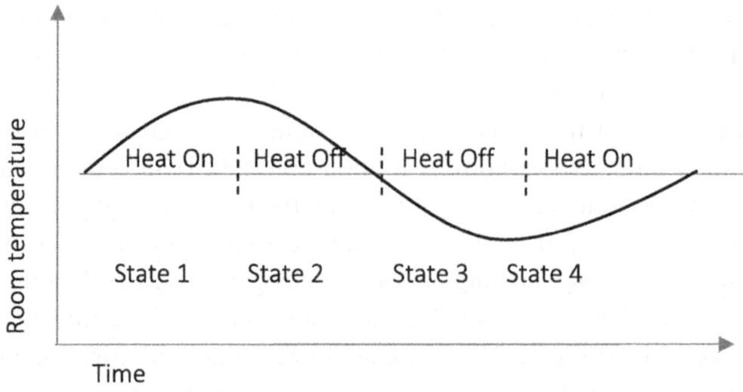

Fig. 6.2 Adding information and logic transforms mechanistic control into a smart system

not, and so forth, providing mechanisms for the use of information for smart system control.

Recently, it has been demonstrated that synaptic plasticity only depends on the activity of pre-synaptic and post-synaptic neurons (Bogacz, 2017) which is consistent with reentrant processes forming the basis for neural node functionality at all levels. For example, at the edge of the system interfacing with the environment and at every level of the hierarchically organized network up to the abstract central functions like the reticular activating system which help the system remain in dynamic equilibrium (Damasio, 1999) or applying the critical cycles per second needed to maintain consciousness as a memory system (Budson et al., 2022).

The smart system processes of efficient information flows and logic result in a tighter, more efficient 'fit for purpose' control surface for the thermodynamic system. In the completed model of the micro level of learning, we build on both the mechanistic and smart systems models and add a proposed natural law that motivates continuous learning in all living systems.

6.4.3 Learning Systems

The learning systems model builds on the physical and the information systems and adds new dimensions of prediction and memory of the dissipative structures of the process. The logic circuit made possible by the smart system now becomes a memory of a longer chain of processes with logic gate sequences, information flows and physical flows bundled into an episode and the whole as well as its parts is used as a comparison signal for prediction. This high-level description has been discussed as a prediction and locator system (Hawkins, 2021; Hawkins & Blakeslee, 2004; Mountcastle, 1997) with supporting evidence from cortical studies, including recent brain network studies of the connectome (Anderson & Barbey, 2023; Sporns, 2011) and research on *prospective cognition* in which a range of memories from highly specific episodic memories to broad semantic memories are creatively reconfigured and reassembled into mental simulations like future scenarios, fantasies, and flights of imagination (Addis & Tanguay, 2022). We assert that this is the space for expansion into the adjacent possible and building mental models required for learning. We can now take up the model from Chap. 5 to add concepts from thermodynamics and complexity to the causal learning cycle (Fig. 6.3).

The Carnot engine, the idealized model of an engine with reversible processes that transform heat into work, is metaphorically apt for the mental model space because the latter appears to be *quasi-reversible*. That is, a mental model can be erased, evolved, redrawn, or overturned, and hypotheses and counterfactuals can be entertained as thought experiments at a small cost of spending time and energy on information transmission. In the mental model space, the dissipative structures may be memories, consisting of objects like episodic memory features, memories with semantic properties, and consciousness itself. The physical world outside of the mind, on the other hand, is primarily *irreversible*, as Prigogine pointed out and Kauffman leveraged to create the motivation for creative expansion of autocatalytic processes and entities into the adjacent possible. Time is a one-way process and events cannot be completely reversed. Concrete things that break take larger amounts of energy to fix, and historical accidents accumulate to form structures (and biases) that persist.

The causal learning cycle interpretation of the Carnot engine is likewise an *irreversible* process that creates and *freezes the structure of experience* over time, to use Kurt Lewin's apt metaphor (Lewin, 1935). The process utilizes stored energy to dissipate the process of sensory and mental inputs into symbols, a process Suzanne Langer called 'symbolization' (Langer, 1942). When the symbols are close reproductions of existing episodic or semantic structures, the learning is closer to reinforcement. When they are more unique, then organizing the data input seems to utilize mechanisms like zero-shot learning (Rezaei & Shahidi, 2020), few-shot imitation (Bhoopchand et al., 2023), and off-policy classification (Irpan et al., 2019)

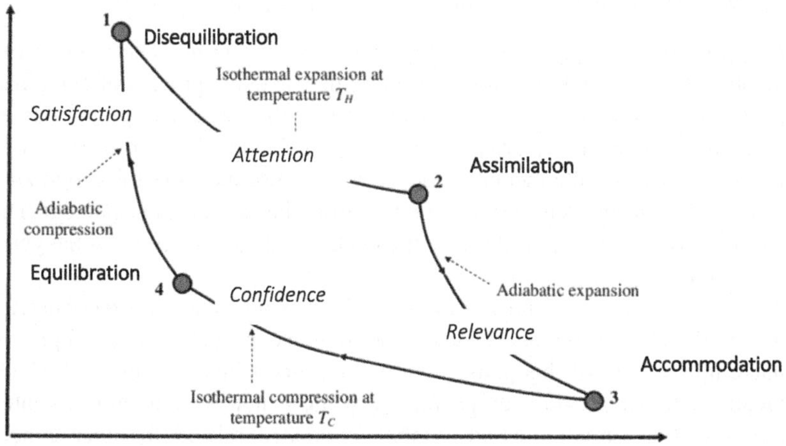

Fig. 6.3 Piaget-Keller-Kauffman (PKK) causal learning cycle

from machine learning and prospective cognition (Addis & Tanguay, 2022) from cognitive neuroscience.

The prefix **Iso** means 'the same everywhere.' In the terms *isothermal expansion* and *isothermal compression* (Fig. 6.3) energy is irreversibly added or removed at the same rate across the boundary of internal and external worlds. Expansion can be understood as *search*, following Thagard (2002) and Kauffman (2010) and compression is understood as *symbolization*, following Langer (1988) and Baum (2003). The two *iso* phases are moments of conserved physical energy doing work in the internal world (e.g. utilizing the in-flow of data and energy during isoexpansion) and the external world (e.g. utilizing the out-flow of data and energy during isocompression).

Adiabatic means 'no heat exchanged', in phases that are primarily insulated from the external world. Note that in the abstract mental model space, a reverberating physical process dominated by information that is isolated from the senses (e.g. episodic and semantic memories) and the reentrant processing of existing concepts (e.g. creative flights of fantasy and simulated reorganization of ideas), adiabatic processes are quasi-reversible internal expansion (search) and compression (symbolization) processes. The two adiabatic phases of the causal learning cycle are therefore moments of non-conserved quasi-reversible processes.

Theories of the mechanisms for these processes at the micro level span across neurology, computational biology, mathematics, and computer science. For example, psychological, cognitive, and brain-based theories of learning can be treated as aggregation or dissipation sub-models (Prigogine, 1976), quasi-homomorphisms (Holland, 1998, 2019), or persistence homologies (Aktas et al., 2019) of the Piaget-Kauffman-Keller model of individual learning. To maintain the narrative flow focused on explaining the broader implications of the three-level hierarchy, notes in Chap. 14 and appendices will be offered rather than delve into these topics in depth now.

Briefly discussing the four Kauffman phases using Piaget's and Keller's theories may help bridge the ideas to traditional learning theory in education.

1. *Disequilibration* brings new information to our *attention*. The volume of the mental model expands because something new is in the process of being recognized, understood, or explained. Over a brief time, the external environmental pressure drops while we are performing internal symbolization work—cognizing new space—in the mental model space because the work of recognizing or symbolizing the new entity robs attention from the external environment. This is the *isothermal expansion* phase of the cycle in which energy is exchanged and conserved.
2. *Assimilation* marks a turning point when the new information is recognized and made *relevant* by using or reconfiguring existing concepts. As internal processes make up the bulk of this phase, like holding one's breath at the end of an in-breath, external pressure drops further while a new or remembered internal representation or compressed symbol is created related to other concepts. External environmental pressure is at its lowest and the expansion of the poten-

tial mental model space is at its maximum. The process shifts from interfacing with environment signals to *searching* existing internal semantic structures, shifting the mechanism to *adiabatic expansion*, which, since isolated from outside sources of energy, can be non-conserved exchanges such as coherence and multiples of pattern-matching processes (Thagard, 2002).

3. *Accommodation* turns a corner of sorts when it increases internal pressure and compresses experience into a symbol. The volume of mental model space contracts, like the initial out-breath, as the symbol relates to and reorders existing symbols. Internal pressure increases as meaning is attached to the symbol and new semantic network connections are made. This phase of *isothermal compression* is physically energetic and therefore follows the laws of the conservation of energy, as stored energy and memories are unfrozen and refrozen (Lewin, 1946) and new network connections are consolidated and *confidence* weights are attached via for example, repetition of neurons firing together resulting in becoming wired together (McCulloch & Pitts, 1943).
4. *Equilibration* is reached when the volume of mental model space returns via *adiabatic compression* (via internal processes somewhat isolated from the external environment) to near the state when disequilibrium first occurred, now slightly expanded with the new or remembered symbolized experience. Internally experienced environmental pressure is at its lowest because of the *satisfaction* of the need to symbolize and maintain the mental model space. Attention then returns to maintaining watch on the ongoing background, the external environment of the real world upheld by current mental models, balanced on the boundary between extant and future externalities.

6.5 Complexity and Educational Theory

The implications and significance of these core ideas have two consequences for our project to guide AI developments that enhance learning processes at the micro, meso, and macro levels. First, the scientific understanding of complexity and evolution in natural systems provides insights into how complex structures and patterns emerge in diverse fields, from chemical reactions to biological systems and social dynamics, and this understanding extends naturally to the three-level hierarchy of distributed intelligence among self, others, and society. Second, because the implications extend beyond physics, to fields like biology, chemistry, ecology, and social sciences, they shape our understanding of dynamic processes across the various disciplines. We have referred to this situation as illustrating the homologous functions of the causal learning cycle within and across the three-level hierarchy.

Drawing links to education, in Table 6.1 are listed concepts stemming from the science of complexity (e.g. nonlinearity, co-evolution, dynamics) brought alongside selected concepts from socio-educational theory and practice (e.g. feedback loops, intersectionality, boundary objects) followed by a brief discussion of the ideas aimed at educators. This table first appeared in (Gibson & Knezek, 2011) without

Table 6.1 Complexity and related concepts for educators

Complex system concept	Brief definition
Nonlinearity	A nonlinear system is one in which the output is not directly proportional to its input; the cause of some response by the system is not the simple sum of the stimuli, as it is in linear systems. This can give rise to surprising, unpredictable behavior.
Feedback loops	Information is recycled, connecting the current state to past states of the system. This cycling is in large part the cause of the nonlinearities, and is also the foundation for growth, learning, and other emergent properties.
Co-evolution through openness	The system accepts "inputs from" and "outputs to" a larger external environment. Systems co-evolve with the niche as well as with other nearby systems through openness.
Self-organization & Adaptation	Impacts on the current state of the system are carried forward into future states of the system. Systems adapt to their environment, including its other systems and dynamics.
Nested relationships & Dynamics	Components of the system may themselves be complex systems and as part of a network of relationships, give rise to trajectories, rhythms, and cycles of activity, rituals, routines and so forth. The dynamics generally lead to three types of overall system behavior ((1) events that rise or expand until exploding or becoming chaotic, (2) events that begin, sustain and die away over time, and (3) cycles of events that sustain for longer periods of time through periodic ups and downs).
Emergent properties	Properties of the whole system depend upon the nonlinear nested relationships of the components and often need a new level of analysis and representation from that of the components. Macro-patterns can emerge. In educational contexts these might be mesosystems of overlapping cultures (family, race, gender, socioeconomic, historical).
Intersectionality	This concept has recently been introduced in the context of complex social systems and represents a particular multi-element dynamic in which sets of elements overlap, combine, collide, and settle into a temporary equilibrium through coherence of their mutually causal relationships.
Boundaries	Since the system has nested relationships, there are boundaries that separate the elements from each other as well as from the larger encompassing environment. In social systems research the idea of "boundary-crossing objects" or "boundary objects" points out that things like concepts, terms, definitions, values, words, etc. that have meaning in one context might take on different meanings after crossing a boundary into another context.

the explicit benefit of explanation of the Piaget, Keller, and Kauffman theories. The discussion of the table has been edited to better fit the context of the current project to guide AI development of learning processes.

Nonlinearity In real systems, nonlinearity is ubiquitous, and is more prevalent than current linear models and much present-day research in educational technology indicates. Educational research traditions persist in allowing linear simplifications to dominate the discourse, even though the world does not behave in a straight line, neither do its learners, classrooms, teachers and communities. The linear worldview feigns that we can understand complex systems by considering them well-behaved and at root, no more complex than a sand pile. Despite the well-accepted fields of knowledge from developmental psychology, classroom observations, educator development, and the social foundations of education, many current educational research approaches are founded on *the assumption that variance in data is itself constant*! Simple linear models are satisfied to view the average or mean of data and assume that the underlying reality can be represented as a line. The observed means and their measures then become crucial building blocks for finding correlations and making inferences. But this worldview needs to introduce time and complexity into the equations and models, as Prigogine and Kauffman point out. Education practitioners need to become comfortable, for example, with new concepts such as hyperlinked learning progressions and 'Internet time' (Bjerede et al., 2010) replacing linear scope and sequence conceptions of knowledge transmission.

Classroom realities are dynamically changing over time, and they occasionally display surprising behavior, such as a student who is learning a new mathematical method, or a group of students experiencing the arc of a classroom's history from the beginning of school to the end of the semester. A better analogy than a line for these kinds of evolving systems is a musical composition, which at times has a single melody, then suddenly breaks into multipart harmony, and may become quiet after a storm of sound. At each point in time in the composition, listeners experience 'frequencies,' but that experience is not represented by the average frequency at a particular moment, and certainly not the mean frequency of the entire piece. Imagine the absurdity of asserting that Beethoven's Ninth Symphony could be summarized as 'an average 256 cycles per second.' Instead, the whole is more than the sum of its parts, and *it needs to be perceived as it is unfolding to comprehend it.*

Feedback Loops Information cycling in a system, or feedback loops, are mechanisms that relate past states to the present. The current state of a student's mind or a classroom's profile is not randomly related to the immediate past state; it is a mixture of reverberations of the state just past, with some newly evolved and additional information arising in the present. This property of feedback often violates an assumption of the ordinary least squares method used in linear regression analysis, which says that error terms are not supposed to be related to each other. So, the typical methodology of some educational research may be unfit to represent the evolution of a real system. This might help explain why *reflection* is often viewed as a qualitative matter, when it is instead, also a highly important matter for a quantitative understanding of the mechanics of dynamic systems. Teacher education programs that have embraced the language of reflection in the qualitative sense, also need to appreciate the significance of information cycling in quantitative models. To remedy this problem, new directions in quantitative and mixed methods such as *quantitative ethnography* (Shaffer, 2017) need to be introduced into and strengthened in educational research.

Openness Openness refers to a function of exchanges of matter or information at the boundary of any system and its environment. This exchange zone is a co-producer of the system's behavior and is inseparable in any analysis that attempts to account for dynamic behavior. Qualitative methods have long appreciated the importance of context for understanding the relationships influencing system evolution. Learners are better understood, for example, when we know their culture, home life, first language, as well as their interests, aspirations and learning preferences. However, in some quantitative models, these aspects are left aside, or they are added in as explanatory descriptions that enhance a primarily linear analysis. In a few rare cases, attempts are made to quantify and integrate exogenous variables into the linear framework, but the models created by traditional linear methods often ignore the dynamic features of nonlinearity, recycling and openness.

Openness in the learning environment is achieved by promoting the exchange of information between students and people outside of the classroom. Teachers need to know how to initiate exchanges and leverage them for student learning, treating the vast open knowledge system as a potential ally. Ubiquitous mobile technologies will quicken the establishment of openness in the educational system (Shaffer, 2017). If educators preserve the right to remain close advisors to their students, then the benefits of openness far outweigh any potential dangers (e.g. from bad information, predators, commercial advertising interests); every external exchange is an opportunity for learning if the teacher is a trusted guide. Benefits to both the teacher and students include the renewal of energy for essential questions, access to higher levels of expertise, a built-in external audience for performance, higher levels of realism and authenticity, and better more up to date information, among many others. Crucial to understanding openness, teachers need to believe that they and their students can make contributions of real value to the outside world, because healthy sustainable exchanges, across the boundary between classroom and outside world, are two-way streets (Gibson, 2000, 2008).

Co-evolution This concept encompasses the idea that simultaneous change and mutual adaptation lead to structural changes, such as consolidating a memory in an individual, or an expert group agreeing on a paradigm shift (Kuhn, 1970), or constructing a new body or capability as a form of species memory. In like fashion, school structure is for the most part, a frozen memory of the cultural transmission model and hierarchical bureaucratic form of education still prevalent today. Like short-term memory becoming long-term through repeated practice, often under time or energy pressures, structural change in education involves establishing new practices in the presence of highly valued incentives to persist with the needed changes.

Co-evolution also links to *collective intelligence* and *distributed cognition* (Jenkins et al., 2009) conceptions of the cultural nature of knowledge. For example, knowledge about the sun is not learned first-hand through experience, but from a shared culture that already knows that fact. The fundamental fact of the co-evolved social construction of knowledge should cause teachers to abandon the notion that knowledge is primarily what is in the student's head and what can be accessed from unaided memory when prompted on a quiz or test. What may be most important is whether a student knows when and where knowledge can be appropriated and assimilated in their search for meaning, understanding, and explanation.

Nested Relationships and Dynamics Hierarchical linear modeling efforts attempt to understand nested relationships and help analyze the layers of educational system policy and practice complexity. Educational researchers need additional training in network analysis to leverage the considerable advances taking place now in biology, earth systems sciences, physics, medicine, and elsewhere. Leadership and policy research and practice have explored a variety of approaches to understanding nested systems (Jenkins et al., 2009).

Emergent Properties The nonlinear worldview implies that averaging over time is inaccurate for some modeling purposes. It also means that schools of education need to (1) employ new research methodologies with nonlinear dynamical approaches and (2) acculturate new researchers with a well-informed and balanced critical awareness of the limitations of linear methods. The new methods may include helpful devices from the traditional statistical toolkit, such as nonparametric statistics, but more importantly, schools of education need to develop new nonlinear quantitative methods and tools of research to join the current toolkits in qualitative and quantitative methods (Gibson & Knezek, 2011).

6.6 Limitations and Conclusions

The chapter's reliance on complex system theories and models to explain educational phenomena does not account for all variables and interactions in real-world educational settings, which involve structures and processes in addition to learning processes (e.g. political biases, historical and economic constraints, structural assumptions). The proposed causal learning cycle model, while innovative, may face challenges in empirical validation due to the inherent complexity and nonlinearity of educational systems as well as the subtlety of measurable signals of learning processes at the micro, meso, and macro levels. Additionally, the application of dissipative structure theory and autocatalysis to learning processes, although insightful, may be limited by the abstract nature of these concepts when translating them into practical educational strategies or policies. The chapter's focus on nonlinearity and system openness also suggests a limitation in predicting outcomes within educational systems, as these characteristics can lead to unpredictable and emergent behaviors that are difficult to accurately model. Lastly, the chapter's interdisciplinary approach, while a strength, may also limit its accessibility to researchers who are not familiar with the complex systems theory or the specific scientific theories it draws upon.

Nevertheless, the chapter aimed to contribute to the field of educational research by applying complex system concepts to understand the structures and dynamics of educational systems. The chapter discussed the openness of learning systems at micro, meso, and macro levels, the co-evolution of those levels with their environment, and the impact of current states on future states, which aligns with the concept of complex adaptive systems. It noted the important yet limited application of hierarchical linear modeling to analyze nested relationships within educational policy and practice, acknowledging the complexity of these systems. The chapter advocated for the necessity of additional training for educational researchers in network analysis, drawing from significant advancements in various scientific fields such as biology, computer science, physics, and medicine.

By integrating nonlinearity, the chapter highlighted the unpredictable nature of educational outcomes and the inadequacy of linear models to fully capture the dynamics of learning environments. Furthermore, the chapter incorporated the principles of dissipative structure theory and autocatalysis to explain how learning processes evolve and self-organize, suggesting a paradigm shift in understanding educational system behavior and evolution. Lastly, it proposed a completed model of the causal learning cycle, influenced by the work of Piaget, Keller, Prigogine, and Kauffman, to describe the complex nature of learning processes.

6.7 Moving Forward

This chapter completed the model of the causal learning cycle analogizing with the features of a control system that integrates flows of conserved energy and non-conserved information during learning processes. It applied Kauffman's logic to the causal learning cycle, drawing from his analogy with the 4 phases of an idealized thermodynamic engine to describe the irreversible process of creating and freezing the structure of experience over time. In the context of educational learning theory, the Kauffman mechanism integrated into the Piaget-Keller learning cycle may help bridge traditional learning theory with complex systems thinking. The conceptual insights gained by adding the Kauffman mechanism to conceptions of learning could potentially lead to new directions in quantitative and mixed methods in educational research.

In the next chapter, the implications of the PKK model are used to envision AI applications that scaffold and support the causal learning cycle in individual learning.

References

Addis, D. R., & Tanguay, A. (2022). *Prospective cognition and its links with memory*. https://doi.org/10.31234/osf.io/w69nd

Aktas, M. E., Akbas, E., & Fatmaoui, A. E. (2019). Persistence homology of networks: Methods and applications. *Applied Network Science, 4*(1). https://doi.org/10.1007/s41109-019-0179-3

Anderson, E. D., & Barbey, A. K. (2023). Investigating cognitive neuroscience theories of human intelligence: A connectome-based predictive modeling approach. *Human Brain Mapping, 44*(4), 1647–1665. https://doi.org/10.1002/hbm.26164

Bandura, A. (2001). Social cognitive theory: An agentic perspective. *Annual Review of Psychology, 52*(1), 1–26. https://doi.org/10.1146/annurev.psych.52.1.1

Baum, E. (2003). *What is thought*. MIT.

Bhoopchand, A., Brownfield, B., Collister, A., Dal Lago, A., Edwards, A., Everett, R., Fréchette, A., Oliveira, Y. G., Hughes, E., Mathewson, K. W., Mendolicchio, P., Pawar, J., Pîslar, M., Platonov, A., Senter, E., Singh, S., Zacherl, A., & Zhang, L. M. (2023). Learning few-shot imitation as cultural transmission. *Nature Communications, 14*(1), 7536. https://doi.org/10.1038/s41467-023-42875-2

Bjerede, M., Atkins, K., & Dede, C. (2010). A special report: Ubiquitous mobile technologies and the transformation of schooling. *Educational Technology, 50*(2), 3–7. JSTOR.

Bogacz, R. (2017). A tutorial on the free-energy framework for modelling perception and learning. *Journal of Mathematical Psychology, 76*, 198–211. https://doi.org/10.1016/j.jmp.2015.11.003

Budson, A. E., Richman, K. A., & Kensinger, E. A. (2022). Consciousness as a memory system. *Cognitive and Behavioral Neurology, 35*(4), 263–297. https://doi.org/10.1097/WNN.0000000000000319

Damasio, A. (1999). *The feeling of what happens: Body and emotion in the making of consciousness* (Issue 4) (pp. 323–326). Harcourt Brace and.

Dinsmore, D. L., Alexander, P. A., & Loughlin, S. M. (2008). Focusing the conceptual lens on metacognition, self-regulation, and self-regulated learning. *Educational Psychology Review, 20*(4), 391–409. https://doi.org/10.1007/s10648-008-9083-6

Edelman, G., & Mountcastle, V. (1982). *The mindful brain*. MIT Press. https://mitpress.mit.edu/9780262550079/the-mindful-brain/

Gibson, D. (2000). Growing towards systemic change: Developing personal learning plans at Montpelier high school. In *In "dynamics of change in high school teaching: A study of innovation in five Vermont professional development schools"*. The Regional Laboratory.

Gibson, D. (2008). Make it a two-way connection: A response to "Connecting informal and formal learning experiences in the age of participatory media". *CITE Journal, 8*(4), n/a.

Gibson, D., & Knezek, G. (2011). Game changers for teacher education. In P. Mishra & M. Koehler (Eds.), *Proceedings of society for information technology & teacher education international conference 2011* (pp. 929–942). AACE.

Hawkins, J. (2021). *A thousand brains: A new theory of intelligence*. Basic Books.

Hawkins, J., & Blakeslee, S. (2004). *On intelligence*. Henry Holt and Company.

Holland, J. (1998). *Emergence: From chaos to order*. Helix Books Addison Wesley/Perseus Books Group.

Holland, J. (2019). Complex adaptive systems: A primer. In D. Krakauer (Ed.), *Worlds hidden in plain sight: Thirty years of complexity thinking at the Santa Fe Institute*. Santa Fe Institute.

Irpan, A., Rao, K., Bousmalis, K., Harris, C., Ibarz, J., & Levine, S. (2019). *Off-policy evaluation via off-policy classification*. https://doi.org/10.48550/ARXIV.1906.01624

Jenkins, H., Purushotma, R., Weigel, M., Clinton, K., & Robison, A. J. (2009). *Confronting the challenges of participatory culture: Media education for the 21st century*. The MIT Press. https://doi.org/10.7551/mitpress/8435.001.0001

Kauffman, S. (2000). *Investigations*. Oxford University Press.

Kauffman, S. (2010). *Reinventing the sacred: A new view of science, reason, and religion*. Basic Books.

Kauffman, S., & McCullogh, W. (1967). *Random nets of formal genes* (Quarterly Progress Report 34; Research Laboratory of Electronics). MIT.

Kuhn, T. (1970). *The structure of scientific revolutions* (2nd ed.). Univeristy of Chicago.

Langer, S. K. (1942). *Philosophy in a new key; a study in the symbolism of reason rite and art*. Harvard University Press.

Langer, S. K. (1988). *Mind: An essay on human feeling*. The Johns Hopkins University Press.

Lewin, K. (1935). *A dynamic theory of personality* (F. Adams & K. Zener, Trans.). McGraw-Hill.

Lewin, K. (1946). Action research and minority problems. *Journal of Social Issues, 2*(4), 34–46. https://doi.org/10.1111/j.1540-4560.1946.tb02295.x

McCulloch, W., & Pitts, W. (1943). A logical calculus of the ideas immanent in nervous activity. *Bulletin of Mathematical Biophysics, 5*, 115–133.

Mountcastle, V. B. (1997). The columnar organization of the neocortex. *Brain: A Journal of Neurology, 120*, 701–722.

Prigogine, I. (1976). Order through fluctuation: Self-organization and social system. In E. Jantsch & C. H. Waddington (Eds.), *Evolution and consciousness: Human systems in transition*. Addison-Wesley.

Rezaei, M., & Shahidi, M. (2020). Zero-shot learning and its applications from autonomous vehicles to COVID-19 diagnosis: A review. *Intelligence-Based Medicine, 3–4*, 100005. https://doi.org/10.1016/j.ibmed.2020.100005

Shaffer, D. (2017). *Quantitative ethnography*. Cathcart Press.

Sporns, O. (2011). *Networks of the brain*. MIT Press.

Thagard, P. (2002). *Coherence in thought and action*. MIT Press.

Chapter 7
AI Roles for Enhancing Individual Learning

7.1 Introduction

In Chap. 6, with the inclusion of the Kauffman mechanism of creative expansion into the adjacent possible, integrated with Piaget's equilibration theory and Keller's theory of motivation, the current model (PKK) asserts that the urge to learn is universal, but toward what end? The PKK model explains how individual psychological states transition into each other in an endless cycle that dissipates energy and information from the environment into internal structures like memories and transformative capabilities, or talents for sensing and making sense, and outputs artifacts that AI might use as indicators of engagement in learning processes. As indicated in the previous chapter, while the aim of creative universal expansion in Kauffman's theory is the emergence of life, intelligence and the survival of species, the current project applies the concept to explain the emergence of agency and self-awareness at the individual level, the emergence of collective knowledge and awareness at the social level, and the emergence of global multiculturalism.

While the latter two themes—the 'adjacent possible' of the social and cultural worlds—will be taken up in later chapters, this chapter will discuss the emergence of agency, self-regulation. and metacognition as the goal of individual learning processes and then purposely apply those ideas to the development of AI that enhances individual learning processes. AI applications built with the specifications and affordances of the causal learning cycle in mind can mediate the cycle (e.g. speed it up, slow it down, pause and restart it, and reinforce it) by splitting attention between the global goal of expanding agency, self-awareness and metacognition of the individual while monitoring and assisting the individual through the nodes and relationships of the learning cycle network.

To illustrate, the next section will analyze the COPES model of self-regulation (Winne & Hadwin, 1998), compare it to the Piaget-Keller-Kauffman cycle and prepare ideas needed to understand AI as a partner in individual learning.

© The Author(s), under exclusive license to Springer Nature Switzerland AG 2024

D. C. Gibson, D. Ifenthaler, *Computational Learning Theories*, Advances in Analytics for Learning and Teaching,
https://doi.org/10.1007/978-3-031-65898-3_7

7.2 Models of Self-Regulated Learning

Panadero reviewed, analyzed, and compared six models of self-regulated learning (Panadero, 2017) in terms of, among other things, how the theorists conceptualized (a) phases and subprocesses, (b) cognition, motivation, and emotion, (c) top-down or bottom-up processes, (d) automaticity, and (e) context, which are summarized in Table 7.1.

The COPES architecture in the Winne & Hadwin model, which stands for Conditions, Operations, Products, Evaluations, and Standards, suggests a dynamic open system where learners continuously monitor and adjust their cognitive strategies. So, their model is closest to the current project and will be analyzed side-by-side with the Piaget-Keller-Kauffman model to point out similarities and highlight any new aspects. Recall that an open system has an external environment that the entity or agent is 'open to,' two internal functions (memory and transformational capacity) and outputs that the agent puts back into the environment. In the COPES model (Fig. 7.1) discussed in (Hadwin et al., 2011) these functions are evident and further discussed below.

Note that task conditions, performance and external evaluations are external to the agent (e.g. the cognitive system) and several entities or processes are represented as internal, including cognitive conditions, cognitive evaluations using standards, monitoring and control of operations and products. The COPES model treats learning as an update to the cognitive conditions, which are recursively updated based on an evaluation of products, which are themselves determined by the monitoring and control of operations that are seeking to satisfy task conditions.

Comparing the COPES model to the Piaget-Keller-Kauffman (PKK) model highlights the unique contributions of the thermodynamic and informational aspects of the new model (Table 7.2). For example, the PKK model predicts that if the task conditions are far outside the zone of proximal development (e.g. are not adjacent possibilities) then the internal processes may not be engaged due to a lack of *attention*. Also, any task condition that is sufficient to trigger attention must influence or cause an *expansion* of the cognitive conditions; if there is no expansion, then there is no gap needing to be filled with new learning or that needs to be validated via performance. Searching the space of memories and extant capabilities to determine relevance according to the PKK model requires reserves of energy and information that limit the control and monitoring of operations and products. Thus, future measures of those reserves monitored by an AI might trigger foundational interventions to better prepare the individual for attending to and maintaining attention on key features of the task.

The PKK model also characterizes the multimodal and multivariate distances or similarities of task to cognitive condition components as a matter of *relevance* (e.g. psychologically perceived relevance in individual learning processes, socially supported relevance in the meso and cultural levels). The PKK predicts there will be a minimum level of relevance needed for an individual to evaluate to what extent the task is novel or has been seen before and whether any accommodation or update is

Table 7.1 Comparing models of self-regulated learning

Theorists	Phases & Subprocesses	Conceptualizations	Top-down/Bottom-Up	Automaticity	Context
Zimmerman (2010)	Forethought, performance, and self-reflection	Cyclical nature, emphasizing self-efficacy and the interaction between personal, behavioral, and environmental factors.	Integrating both top-down and bottom-up processes through triadic components that interact in cycles of feedback, suggesting that cognitive processes and behavior/ environment influence each other bidirectionally	Automaticity as a developmental achievement within srl that can enhance efficiency.	The learning environment's influence on self-efficacy and the cyclical phases of srl, suggesting that context shapes the personal, behavioral, and environmental aspects of learning
Boekaerts and Corno (2005)	Identification, interpretation, primary and secondary appraisal, goal setting, goal striving, and performance feedback	Dual-processing model focuses on the interplay between learning and well-being, suggesting that learners navigate between a mastery pathway and a well-being pathway.	Implies a top-down approach with its emphasis on personal beliefs and mental processes.	Acknowledges both deliberate and automatic pathways in the regulation of learning, which can influence students' engagement with tasks	The learner's interpretation of the learning task and context, activating different regulatory actions based on the perceived goals, which could be mastery-oriented or well-being-oriented, thus showing a dual pathway influenced by context
Winne and Hadwin (1998)	Open process, with recursive phases: Task definition, Goal setting and planning, Enacting strategies, Adaptation	Four-phase feedback loop	Rejects the dichotomy, emphasizing the importance of cycles of information flow that are both internal to the person and cross the boundary between person and environment, thus presenting a more dynamic and reciprocal interaction.	Model incorporates internal feedback loops and the adaptation of strategies, which could imply a level of automaticity in the regulation of learning once strategies become well-practiced and efficient.	The conditions that learners set for themselves in the initial phase of their model, which then influences the strategies and standards they use for monitoring and controlling their learning

(continued)

Table 7.1 (continued)

Theorists	Phases & Subprocesses	Conceptualizations	Top-down/Bottom-Up	Automaticity	Context
Pintrich (2000)	Forethought, planning, activation, monitoring, control, reaction, and reflection	Integrates motivation and cognition, with a particular emphasis on goal orientation and its impact on learning strategies.	Being goal-driven, could be interpreted as having a top-down orientation, with motivation stemming from personal characteristics; however, it also includes goal orientation, which implicates performance and avoidance goals, linking to boekaerts' well-being pathway and suggesting a more nuanced tdbu interpretation	Accessing prior knowledge, which can occur without conscious thought.	Relates to the role of goal orientation in srl, where the learning context can influence the type of goals students set and the strategies they employ
Efklides (2011)	Task representation and cognitive processing during performance	Interplay between cognition, metacognition, and affect	Interaction among metacognition, motivation, and affect at both macro and micro levels, which suggests a more integrated approach to tdbu	Involves both conscious analytic processes and non-conscious automatic processes, indicating an understanding of srl that encompasses both explicit and implicit knowledge and their effects on behavior	Implies that context influences the person's perceptions of task demands and thus affects the regulation of cognition, motivation, and affect
Hadwin et al. (2011)	Planning, monitoring, control, and regulating phases	Socially shared self-regulation of learning	Expands the concept of regulation to the social context, where groups collectively engage in regulatory activities	Aligned with the notion of implicit regulation in collaborative settings, where group members may engage in regulatory actions that are not fully explicit or conscious	Extends the concept of context to include the social environment, where the regulatory activities are not only individual but also shared among group members, indicating that context includes the collaborative setting and the interactions within it

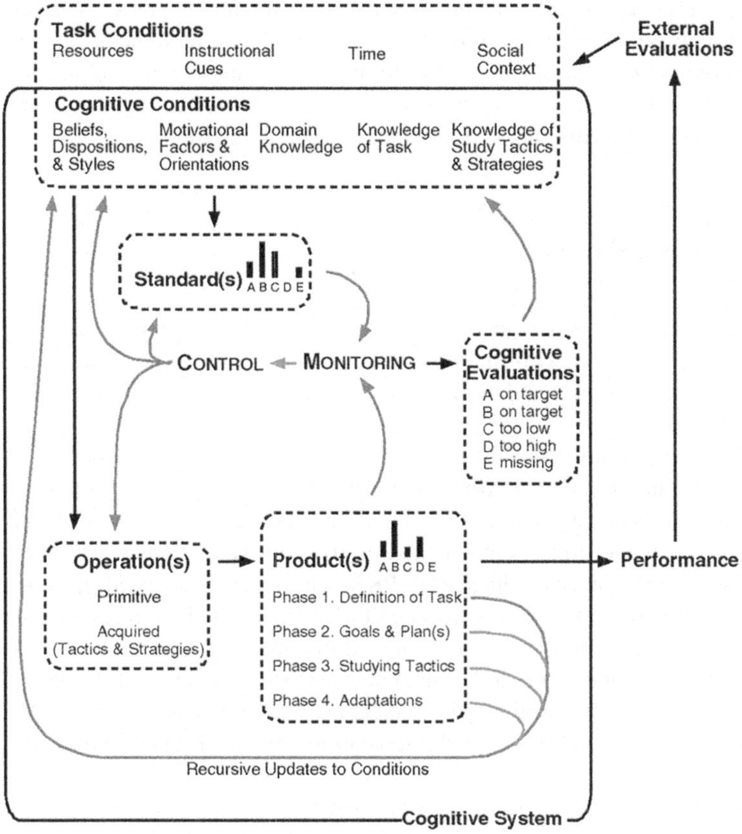

Fig. 7.1 COPES model of self-regulated learning

Table 7.2 Comparing Piaget-Keller-Kauffman and COPES models

Piaget-Keller-Kauffman state & transition	COPES components
Disequilibrium & Attention	External task conditions & External evaluations
Assimilation & Relevance	Cognitive conditions & Cognitive evaluations
Accommodation & Confidence	Operations, Products and Recursive loop to Cognitive Conditions, with updates as needed
Equilibration & Satisfaction	Control, Monitoring, Standards leading to Products shared externally

needed to initiate operations and production of products. To continuously evaluate relevance, the PKK model implies that an internal confidence standard is engaged related to each of the multivariate cognitive conditions, which are distinct from external evaluation criteria for the task. Future AI interventions might manipulate either task conditions or inform the individual about one's cognitive conditions to increase relevance in either external or internal evaluation cycles. These comparisons and observations concern the expansion phase (e.g. both the iso and adiabatic expansions) of the learning cycle and point to information that could become available to the individual with appropriate AI that could assist in triggering accommodation, confidence building, and satisfaction via equilibration.

7.3 AI as a Partner

This section considers the question of how to develop artificial intelligence applications that can partner with the individual to enhance the Piaget-Keller-Kauffman model's network nodes and connections and thereby enhance individual learning processes.

AI as a partner in individual learning can be understood as an integration of human and automated cognition, where the automated portion possesses a high degree of independent agency and is thus able to be a critical friend for inquiry into, and action at, the leverage points of the Piaget-Keller-Kauffman cycle. Interventions and new information can influence the PKK learning cycle to speed-up, slow-down, and in other ways mitigate and mediate the four learning process stages and related causal connecting links. With such potential in mind, four broad roles for AI can be aligned with the network's nodes and edges to help an individual through the processes and stages of learning: (1) Personalized recommendations, (2) Individual learning analytics, (3) Pedagogical interventions, and (4) Formative and summative feedback.

The goals of AI applications aimed at enhancing individual learning can be sensitized to and activate conditions that help drive the Piaget-Keller-Kauffman system into the next state or transition, to 'speed up' or in other ways attenuate and mediate the cycle of learning. For example, if an individual is in a state of *disequilibrium*, then the most probable related outcome is that the individual's *attention* will be drawn to what initiated the disequilibrium. Following signals of attention like eye gaze, time on task, expressed thoughts, and interactional behaviors, could lead the AI to infer what the learner needs in the moment to transition to the state of *assimilation*, where *relevance* is either found or created by the individual and without which, the rest of the learning cycle is disengaged or ignored (Treisman, 1964). If the individual remains in the state of *accommodation*, *confidence* can be created and enhanced via practicing new knowledge and skills, extending the range and scope of current expertise and broadening the base of expertise, which has been shown to increase the contextual recovery of memories as well as strengthen the long-term memory of the appropriate contextual superposition of a representational network assembly (Li et al., 2016). In the state of *equilibration*, some sense of *satisfaction* in monitoring the environment or achieving internal goals has been reached. These phases and focus areas for AI development at the micro level are listed in Table 7.3 and discussed further below.

Table 7.3 The micro-level model of individual learning: Piaget-Kauffman-Keller Theory enhanced by AI

Phases or states of Piaget-Keller-Kauffman nodes and edges	The focus of AI systems co-participating in learning processes
Disequilibration—iso expansion *causes or mediates* **Attention**	**Personalized recommendations**
Assimilation—iso compression *causes or mediates* **Relevance**	**Individual learning analytics**
Accommodation—adiabatic expansion *causes or mediates* **Confidence**	**Pedagogical interventions**
Equilibration—adiabatic compression *causes or mediates* **Satisfaction**	**Formative and summative feedback**

Disequilibration AI applications can make personalized recommendations that stretch the learner's current boundaries of knowledge and capability and suggest the next best strategic moves by applying machine learning to user attributes and activity data (Pea, 2014). AI can bring up a new idea or observation, drawing attention to something not noticed before, or the learner might prompt the AI to look for and then share something new to consider (Montaner et al., 2003). Related examples include an 'alerting dashboard' for improving scientific inquiry (Dickler et al., 2021) and research on knowledge tracing of attention (Rodrigues et al., 2022).

Assimilation During the assimilation phase of integrating something new into what was already known, AI can provide information for analysis by learners to help them meet their needs—*learner-centered learning analytics*—allowing someone to see for example where they stand compared to others or their past and making clear the current goals and paths forward (Shum et al., 2019). For example, AI might detect deep versus surface learning strategies used by the learner (Gasevic et al., 2017). Another approach of AI assisting assimilation and future accommodation is providing visualized model-based feedback, in which concept maps are offered that are structurally and semantically like expert solutions (Ifenthaler, 2011).

Accommodation Building confidence in what one knows requires accommodating new knowledge into existing habits, skills, and practices often assisted by tutors, which today can be offered automatically by empathetic pedagogical agents (Sabourin et al., 2011). Historically, research on such intelligent tutors has envisioned selecting a teachable moment and influencing the learner at that moment (Shute & Psotka, 1994). Recently, improvements in AI interactions can also flag the urgency of instructional interventions with promise for automated tutors (Yu et al., 2021). Other *pedagogical interventions* support study success (Ifenthaler et al., 2019) and provide data and evidence from virtual practicums, games, and simulations for learning (Gibson & Jakl, 2013). Finally, aggregating from a common learner model across multiple learners opens an influence channel to the meso level, for example, by informing a dynamic version of instructional design (Lockyer & Dawson, 2012).

Equilibration Finally, in the phase of return to equilibrium, making AI explainable via *feedback to the learner* is seen as key to making lasting impacts (Khosravi et al., 2022) since the learner's understanding of feedback is key to taking informed action and next steps. AI can also accumulate evidence over time and create linkages and analyses concerning standards and outcomes of interest, including by using advanced natural language processing along with deep learning techniques to interpret texts, images, and transcripts to infer levels of knowledge and capability.

7.4 Elaborating the AI Roles of Individual Learning Processes

In the following section, additional research literature sources are aligned with the four broad roles for AI, to aide in thinking about the possibilities for AI application development. The statements might be thought of as 'what is known about these topics' and while the list is not intended to be comprehensive, the hope is that by presenting them in a holistic framework with additional detail supporting the assumption of causation propelling individual learning into and out of the states and transitions as outlined, these elaborations may inspire new ideas and cross-connections across the domains involving constructs for achieving the goal of creating AI applications that are sensitive to and effectively intervene as well as scaffold the learning cycle.

7.4.1 *Personalized Recommendations*

1. **Adaptive Learning Paths.** AI can enable the creation of personalized learning paths based on individual strengths, weaknesses, and learning styles and can continuously monitor and balance the goals for each basis (Shute & Zapata-Rivera, 2012; Siemens & Baker, 2012).
2. **Content Customization:** AI can tailor educational content, resources, and exercises to suit learners' specific needs and preferences (Brusilovsky, 2001; Ifenthaler et al., 2017).
3. **Dynamic Feedback Loops:** AI can continually adjust recommendations based on the learner's evidence of understanding and acting on feedback as well as on real-time performance data (Amazona & Hernandez, 2019; Ifenthaler, 2011).

7.4.2 Learning Analytics

1. **Data-Driven Insights:** AI can process vast amounts of learning data to identify patterns, trends, and areas for improvement (Baker & Siemens, 2014).
2. **Predictive Analytics:** AI algorithms can forecast learner behavior and performance, aiding in proactive intervention strategies (Gasevic et al., 2017; Waheed et al., 2022).
3. **Educational Impact:** Learning analytics derived from AI can inform ethical decision-making, curriculum design, and resource allocation (Ifenthaler et al., 2017; Pardo & Siemens, 2014; Shum et al., 2019).

7.4.3 Pedagogical Interventions

1. **Adaptive Teaching Strategies:** AI can support educators in adapting teaching methods to suit individual learner needs and preferences (Ifenthaler, 2010).
2. **AI-Supported Instruction:** AI can provide real-time assistance, supplemental materials, and interactive tools for educators (Jones & Castellano, 2018).
3. **Teacher-Student Interaction Enhancement:** AI can foster stronger teacher-student relationships by providing insights and strategies to address individual learning challenges (Christensen et al., 2022).

7.4.4 Formative and Summative Feedback

1. **Timely Assessments:** AI can facilitate timely and detailed assessments of learner progress, aiding in immediate feedback (Bennett, 2022).
2. **Individualized Feedback:** AI can generate personalized feedback addressing specific strengths and areas for improvement (Dai et al., 2023).
3. **Feedback Integration:** AI can ethically integrate formative and summative feedback to create a comprehensive understanding of learner performance (de Freitas et al., 2017; Dieterle et al., 2022; Khosravi et al., 2022).

7.5 Evidence and Indicators from Historical Learning Theories

Other 'cycle of learning' models have been proposed throughout history including (Dewey, 1933; Köhler, 1969; Kolb & Fry, 1975; Lewin, 1946; Mumford, 1992). With a degree of overlap and comparable functions with the current project (Table 7.4) these models have made significant contributions to learning theory.

Table 7.4 Comparison of cycle of learning theoretical frameworks

Kohler	Dewey	Lewin	Kolb & Fry	Mumford
Preparation	Suggestions	Plan		Planning to apply or test
Incubation	Intellectualization	Action	Concrete experience	Doing something
Insight	Hypothesizing		Formation of abstract concepts	Reflecting on the experience
	Mental Elaboration		Observation and reflection on experience	Concluding, developing a theory
Verification	Testing the Hypothesis	Fact Finding	Testing new concepts	

Without attempting to undertake a deep dive into each theory, and accepting some risk of oversimplification, a brief comparison can help make it clear that the idea of *stages* (represented in the current project as *nodes* of a network) and a progression of *processes* (represented here as the *causal link edges* that form a directed loop of the cycle) have been historically proposed as elements of various learning theories. Against this backdrop, the current project has introduced a new thermodynamic information theory explanation of the nodes and links, following from complexity and computational biology theories.

A closer look at the historical theories produces evidence of *homologous functions*, often implicit in the historical theories with neither reference to a mechanistic or causal explanation nor an assertion of generality across micro, meso and macro levels. For example, Wolfgang Köhler's work in "The Task of Gestalt Psychology" (Köhler, 1969) touches upon the concepts of preparation, incubation, insight, and verification in the context of problem-solving and the nature of creative thought, which are similar to the states in the Piaget-Keller-Kauffman model of the learning cycle.

1. **Preparation:** In the initial stage of problem-solving an individual prepares by gathering relevant information and understanding the components of a problem. This stage involves becoming familiar with the elements that require a solution. Note the similarity to the state of disequilibrium and becoming attentive to something currently unresolved.
2. **Incubation:** During this period the problem is not actively or consciously worked on. Instead, the mind continues to process the information subconsciously. Köhler suggests that during this incubation period, there may be unconscious cognitive processes at work, leading to novel insights. Note the similarity to the state of assimilation, where current sensory and memory resources are searched for similarities and closeness.
3. **Insight:** Insight is then a sudden and profound understanding or solution to a problem. It is characterized by a restructuring of the mental representation of the problem. Note the similarity to the transition and then the moment of accommodation when there is an active re-making or adaptation of the mental models closely linked to the current situation.

4. **Verification:** After the insight has occurred, there is a stage of verification that involves testing whether the newfound solution effectively solves the problem and aligns with the information gathered during the preparation stage. Note the similarity to the transition via an assessment of satisfaction leading to the return to a state of dynamic equilibration.

Here, in this comparison to Köhler is an analytic process that could be repeated from Dewey to Mumford. The contribution of the thermodynamic and information system causal forces outlined in Chaps. 1, 2, 3, 4, 5, and 6 has added concrete mechanisms for the proposed state transitions and relationships while preserving the historical theories.

7.6 Limitations and Conclusions

The evidence from historical learning theories, while foundational and influential, may not provide sufficient empirical proof in the context of modern educational technology and AI applications. The chapter acknowledges the contributions of various learning theories that define stages and a progression of processes in learning that are analogous to the nodes and causal link edges in the current project's learning cycle model. But more empirical evidence is needed, perhaps from design-based research, robotics and AI in education research, to support the new theory in its thermodynamic and information-based causal aspects.

Moreover, while the chapter draws parallels between historical theories and the Piaget-Keller-Kauffman model, the introduction of a thermodynamic information theory explanation of the nodes and links is a departure from the empirical foundations of the original theories. The theoretical integration may be innovative but needs direct tests and empirical evidence to support its efficacy in enhancing learning through AI. Additional empirical validation through research data is necessary to confirm the new model's effectiveness and to address any potential limitations or ethical considerations.

The chapter relies primarily on previously published studies to substantiate the components of the theory as well as the components' potential effectiveness in the proposed AI roles promoting learning processes. Significant gaps in knowledge are remaining if the aim is to influence the development of AI applications that promote learning processes. The chapter also has not addressed the full range of potential challenges or ethical considerations associated with the implementation of AI in educational settings, which is crucial given the sensitive nature of social learning environments. The focus on learning processes rather than broadly on education and its structures and politics, helps ameliorate this limitation, but more discussion is needed and would be welcome as the theory is extended into the social and cultural realms in later chapters.

Additionally, while the chapters concerning the micro level refer to several important historical theories, they do not deeply analyze each one, which may have resulted in an oversimplification of some complex educational constructs. The absence of a critical evaluation of the limitations of AI in learning thus far in the narrative, for example, such as algorithmic biases or the digital divide, suggests a need for a more comprehensive analysis of the subject matter, even if limited to 'learning processes' and held distinct and apart from the structural, political, and humanitarian aims of education.

Nevertheless, despite its limitations, the chapter completes the proposal of a new testable model of learning processes at the micro level, with leverage points highlighted where AI can play significant roles in enhancing individual learning processes, grounded in the specific functions in the Piaget-Keller-Kauffman learning cycle. The PKK model describes a universal mechanism for the emergence of agency and self-regulation in learning as a natural evolution of intelligence, via a process of dissipating energy and information in increasingly efficient ways while expanding upon and elaborating an essential four-stage causal cycle of learning processes. The use of AI by an individual to enhance learning is an extension of the same process of distribution of intelligence that gave rise to language, the arts, and other cultural and historical processes.

This chapter suggested that if developed with the PKK framework, AI can act as an effective partner in learning, offering personalized recommendations, individual learning analytics, pedagogical interventions, and formative and summative feedback to facilitate the learning stages of disequilibration, assimilation, accommodation, and equilibration. In support of this idea, the chapter drew on historical learning theories, aligning them with the proposed model to demonstrate the continuity of learning theory evolution and the potential for AI to provide concrete mechanisms for state transitions within the learning cycle.

7.7 Moving Forward

Chapters 4, 5, 6, and this chapter have outlined the micro level of the book's 3-level learning theory framework and discussed how current literature and historical theories integrate into the new model of individual learning processes. The model has introduced new causal mechanisms based on similarities in computational and neurological functions that are founded in a thermodynamic information theory to explain individual learning processes. In Chap. 8, the narrative turns to the social dimension to begin to apply homologous functional concepts to the expanded context of self and others.

References

Amazona, M. V., & Hernandez, A. A. (2019). Modelling student performance using data mining techniques: Inputs for academic program development. In *Proceedings of the 2019 5th international conference on computing and data engineering* (pp. 36–40). https://doi.org/10.1145/3330530.3330544

Baker, R., & Siemens, G. (2014). Educational data mining and learning analytics. In R. K. Sawyer (Ed.), *The Cambridge handbook of the learning sciences* (2nd ed., pp. 253–272). Cambridge University Press. https://doi.org/10.1017/CBO9781139519526.016

Bennett, R. (2022). AI and equity in assessment and learning. *Admissions Testing Research, 1*, 7–15.

Boekaerts, M., & Corno, L. (2005). Self-regulation in the classroom: A perspective on assessment and intervention. *Applied Psychology: An International Review, 54*(2), 199–231.

Brusilovsky, P. (2001). Adaptive hypermedia. *User Modeling and User-Adapted Interaction, 11*(1/2), 87–110. https://doi.org/10.1023/A:1011143116306

Christensen, R., Hillaire, G., Marvez, G., Azukas, M., Holben, D., Gregory, S., & Ledger, S. (2022). Using virtual simulations in teacher education to develop equitable teaching strategies. In E. Langran (Ed.), *Proceedings of society for information technology & teacher education international conference* (pp. 484–488). AACE.

Dai, W., Lin, J., Jin, F., Li, T., Tsai, Y.-S., Gasevic, D., & Chen, G. (2023). *Can large language models provide feedback to students? A case study on ChatGPT* [Preprint]. EdArXiv. https://doi.org/10.35542/osf.io/hcgzj

de Freitas, S., Gibson, D., Alvarez, V., Irving, L., Charleer, S., & Verbert, K. (2017). How to use gamified dashboards and learning analytics for providing immediate student feedback and performance tracking in higher education. *WWW 2017*, 6.

Dewey, J. (1933). *How we think: A restatement of the relation of reflective thinking to the educative process*. D.C. Heath and Company.

Dickler, R., Gobert, J., & Sao Pedro, M. (2021). Using innovative methods to explore the potential of an alerting dashboard for science inquiry. *Journal of Learning Analytics, 8*(2), 105–122. https://doi.org/10.18608/jla.2021.7153

Dieterle, E., Dede, C., & Walker, M. (2022). The cyclical ethical effects of using artificial intelligence in education. *AI & Society*. https://doi.org/10.1007/S00146-022-01497-W

Efklides, A. (2011). Interactions of metacognition with motivation and affect in self-regulated learning: The MASRL model. *Educational Psychologist, 46*(1), 6–25. https://doi.org/10.1080/00461520.2011.538645

Gasevic, D., Jovanovic, J., Pardo, A., & Dawson, S. (2017). Detecting learning strategies with analytics: Links with self-reported measures and academic performance. *Journal of Learning Analytics, 4*(2), 113–128. https://doi.org/10.18608/jla.2017.42.10

Gibson, D., & Jakl, P. (2013). *Data challenges of leveraging a simulation to assess learning* (p. 8). Pragmatic Solutions.

Hadwin, A., Järvelä, S., & Miller, M. (2011). Self-regulated, co-regulated, and socially shared regulation of learning. In B. Zimmerman & D. Schunk (Eds.), *Handbook of self-regulation of learning and performance* (pp. 65–84). Routledge/Taylor & Francis.

Ifenthaler, D. (2010). Learning and instruction in the digital age. *Learning, 3–10*. https://doi.org/10.1007/978-1-4419-1551-1

Ifenthaler, D. (2011). Bridging the gap between expert-novice differences: The model-based feedback approach. *Journal of Research on Technology in Education, 43*(2), 103–117.

Ifenthaler, D., Gibson, D., & Dobozy, E. (2017). The synergistic and dynamic relationship between learning design and learning analytics. *ASCILITE, 2017*, 1–5.

Ifenthaler, D., Mah, D., & Yau, J. (2019). Utilising learning analytics for study success: Reflections on current empirical findings. In *Utilizing learning analytics to support study success* (pp. 27–36). Springer International Publishing. https://doi.org/10.1007/978-3-319-64792-0_2

Jones, A., & Castellano, G. (2018). Adaptive robotic tutors that support self-regulated learning: A longer-term investigation with primary school children. *International Journal of Social Robotics, 10*(3), 357–370. https://doi.org/10.1007/s12369-017-0458-z

Khosravi, H., Shum, S. B., Chen, G., Conati, C., Gasevic, D., Kay, J., Knight, S., Martinez-Maldonado, R., Sadiq, S., & Tsai, Y.-S. (2022). Explainable artificial intelligence in education. *Computers and Education: Artificial Intelligence*, 100074. https://doi.org/10.1016/J.CAEAI.2022.100074

Köhler, W. (1969). *The task of gestalt psychology*. Princeton University Press.

Kolb, D., & Fry, R. (1975). Towards an applied theory of experiential learning. In Cooper (Ed.), *Theories of group processes* (pp. 33–38). Wiley.

Lewin, K. (1946). Action research and minority problems. *Journal of Social Issues, 2*(4), 34–46. https://doi.org/10.1111/j.1540-4560.1946.tb02295.x
Li, M., Liu, J., & Tsien, J. Z. (2016). Theory of connectivity: Nature and nurture of cell assemblies and cognitive computation. *Frontiers in Neural Circuits, 10*(April), 1–8. https://doi.org/10.3389/fncir.2016.00034
Lockyer, L., & Dawson, S. (2012). *Learning designs and learning analytics*. https://doi.org/10.1145/2090116.2090140
Montaner, M., López, B., & De La Rosa, J. L. (2003). A taxonomy of recommender agents on the internet. *Artificial Intelligence Review, 19*(4), 285–330. https://doi.org/10.1023/A:1022850703159
Mumford, A. (1992). Individual and organizational learning: The pursuit of change. *Management Decision, 30*(6), 143.
Panadero, E. (2017). A review of self-regulated learning: Six models and four directions for research. *Frontiers in Psychology, 8*(APR). https://doi.org/10.3389/fpsyg.2017.00422
Pardo, A., & Siemens, G. (2014). Ethical and privacy principles for learning analytics. *British Journal of Educational Technology, 45*(3), 438–450. https://doi.org/10.1111/bjet.12152
Pea, R. (2014). *A report on building the field of learning analytics for personalized learning at scale* (pp. 1–24). Stanford University. https://ed.stanford.edu/news/stanford-professor-spurs-movement-build-new-field-learninganalytics
Pintrich, P. R. (2000). The role of goal orientation in self-regulated learning. In *Handbook of self-regulation* (pp. 451–502). Elsevier. https://doi.org/10.1016/B978-012109890-2/50043-3
Rodrigues, T. B. S. F., Souza, J. F. de, Bernardino, H. S., & Baker, R. S. (2022). Towards interpretability of attention-based knowledge tracing models. *Anais Do Simpósio Brasileiro de Informática Na Educação*, 810–821. https://doi.org/10.5753/SBIE.2022.224685
Sabourin, J., Mott, B., & Lester, J. (2011). Computational models of affect and empathy for pedagogical virtual agents. In *Standards in emotion modeling*. Lorentz Center International Center for workshops in the Sciences.
Shum, S. B., Ferguson, R., & Martinez-maldonado, R. (2019). Human-centred learning analytics. *Journal of Learning Analytics, 6*(2), 1–9. https://doi.org/10.18608/jla.2019.62.1
Shute, V., & Psotka, J. (1994). *Intelligent tutoring systems: Past, present, and future* (pp. 2–52). Human resources directorate manpower and personnel research division.
Shute, V. J., & Zapata-Rivera, D. (2012). Adaptive educational systems. In P. J. Durlach & A. M. Lesgold (Eds.), *Adaptive technologies for training and education* (1st ed., pp. 7–27). Cambridge University Press. https://doi.org/10.1017/CBO9781139049580.004
Siemens, G., & Baker, R. S. J. D. (2012). Learning analytics and educational data mining. In *Proceedings of the 2nd international conference on learning analytics and knowledge – LAK '12*. https://doi.org/10.1145/2330601.2330661
Treisman, A. (1964). Monitoring and storage of irrelevant messages in selective attention. *Journal of Verbal Learning and Verbal Behavior, 3*(6), 449–459. https://doi.org/10.1016/S0022-5371(64)80015-3
Waheed, H., Hassan, S.-U., Nawaz, R., Aljohani, N. R., Chen, G., & Gasevic, D. (2022). Early prediction of learners at risk in self-paced education: A neural network approach. *Expert Systems with Applications*, 118868. https://doi.org/10.1016/j.eswa.2022.118868
Winne, P., & Hadwin, A. (1998). Studying as self-regulated learning. In D. Hacker, J. Dunlosky, & A. Graessr (Eds.), *Metacognition in educational theory and practice* (pp. 277–304). Lawrence Erlbaum.
Yu, J., Alrajhi, L., Harit, A., Sun, Z., Cristea, A. I., & Shi, L. (2021). *Exploring Bayesian deep learning for urgent instructor intervention need in MOOC forums* (pp. 78–90). https://doi.org/10.1007/978-3-030-80421-3_10

Chapter 8
Informal Social Learning

8.1 Introduction

In Chaps. 4, 5, 6, and 7 discussing the micro level, the aim was to guide the development of AI to enhance individual learning processes. Such an AI, structured by the four phases and connections of the Piaget-Keller-Kauffman (PKK) model, would leverage the learning cycle mechanisms to increase agency, self-awareness and self-regulation of learning. In this and subsequent chapters concerning the meso level, a social learning theory is proposed that builds upon and embeds the PKK as a subnet in an encompassing network and extends the model to account for the shared processes of social learning. In this chapter the ideas of John Dewey are highlighted, including social responsibility, democracy and real-world learning as a context for individual learning processes. Then, in Chap. 9 the ideas of John Bransford and colleagues will be added, bringing a review of decades of cognitive science research in social settings and the meso level model will then be completed in Chap. 10, devoted to suggesting AI applications aimed at assisting a team or community of learners via computer supported collaborative learning.

8.2 Dewey's Social Pragmaticism

John Dewey was a prominent American philosopher, psychologist, and educational reformer, known for his influential ideas in pragmatism and progressive education. Born in 1859, he greatly impacted education through a philosophy centered on the belief that knowledge should be practical and directly applicable to real-life experiences. He emphasized the importance of learning through experience, critical thinking, and problem-solving rather than rote memorization. Dewey advocated for

© The Author(s), under exclusive license to Springer Nature Switzerland AG 2024
D. C. Gibson, D. Ifenthaler, *Computational Learning Theories*, Advances in Analytics for Learning and Teaching,
https://doi.org/10.1007/978-3-031-65898-3_8

hands-on learning, where students actively engage with the material and their environment.

Dewey's ideas led to the development of progressive education, which focused on a student-centered approach to learning. He believed that education should be tailored to the individual needs of students and should prepare them for active participation in a democratic society. Dewey also contributed to fields beyond education and philosophy, including psychology, ethics, and social and political theory. His ideas about democracy, ethics, and the role of education in society continue to influence discussions and practices in these areas.

In his book 'Experience and Education' (Dewey, 1998) he chastised and reminded progressive educators that his core ideas were based in social science rather than a permissive and seemingly aimless perspective on education, as some progressives had developed. To reiterate the scientific foundations, he posed the question of how it is that a child, before ever being formally taught in school, arrives at school knowing the language and a great many facts about self, family and the world. His answer was the real-world social context in which the child had solved the problems of eating, caring for oneself, and communicating needs. Dewey thus placed significant emphasis on the informal social context and social dynamics of learning, considering them fundamental to the educational process. He believed that learning wasn't just an individual endeavor, but a social and communal activity deeply connected to one's environment and interactions with others.

For Dewey, social interaction was crucial for learning because it allowed individuals to engage with diverse perspectives, collaborate on problem-solving, and develop important social skills. He advocated for an education system that encouraged cooperation, communication, and shared experiences among students. In Dewey's view, the social environment of a classroom or community provides a rich context for learning. Through interactions with peers, students not only gain knowledge but also learn important social and emotional skills like teamwork, empathy, conflict resolution, and communication. These social skills were considered essential for active participation in a democratic society.

Furthermore, Dewey believed that society itself acted as a classroom, providing continuous opportunities for learning through everyday experiences. He emphasized the integration of real-life experiences and societal engagement into the educational curriculum, asserting that students learn best when education is connected to and relevant within their social and cultural context.

8.3 Relating Dewey's Social Learning Ideas to the Micro Level

How do Dewey's ideas interact with or fulfill the phases of the PKK learning cycle in a social learning setting? To consider this question requires a two-sided analysis, first from an individual perspective and second from a team or learning community perspective, considering each of the PKK learning cycle functions:

1. Disequilibration phase leading to Attention—an external event triggers arousal of noticeable differences for an individual as well as a team or organization.
2. Assimilation phase leading to Relevance—an internal information search expands the internal representation space and linkages and identifies meaningful gaps. How this phase appears in both individual and group contexts.
3. Accommodation phase leading to confidence—new symbols and relationships are created to fill the gaps and when externalized, the environment then provides feedback concerning the updated internal models (e.g. knowledge, skills, know-how, expertise).
4. Equilibration phase leading to satisfaction—internal attractors (e.g. set points, cycles and assemblies) are reset to a new normal of monitoring via ongoing cycles of prediction and validation.

To illustrate, the following paragraphs note some of the potential impacts of Dewey's ideas on social learning and connect these impacts to other historical literature.

8.3.1 Disequilibration

Disequilibration in a learning community, in contrast to individual learning, triggers public awareness of a conflict, challenge, or other issue needing political resolution. Experiential learning, for example, may result in conflicts of communication and understanding among group members by introducing new perspectives on experiences and challenges. Collaboration may introduce social challenges and perspectives, potentially disrupting shared cognitive structures such as biases and norms. Personalizing learning may potentially disrupt traditional, uniform educational approaches, and integrating formal and informal learning may disrupt the compartmentalization of knowledge. Continuous learning in the community requires openness concerning intellectual challenges and exploratory ideas, which disequilibrate the status quo. Cognitive conflicts among the group's members may also be created when facing nonroutine, complex problems. The effect of any of these disequilibration events causes the group to turn attention away from standard practices and toward newly arising issues.

Several additional sources in the literature on teams and organizations support the idea of the disequilibration of a shared mental model of a team or organization. For example, the stage theory of small group development (Tuckman, 1965; Tuckman & Jensen, 1977) includes the concept of 'storming' representing that conflicts among members concerning a team task, problem, or challenge are involved in team evolution and need to be settled before 'norming' and 'performing' phases of team development. Edgar Schein's work in organizational culture and leadership (Schein, 2010) provides insights into group processes within teams concerning how groups navigate challenges. Research on teaming and psychological safety (Grand et al., 2016; Hackman & Wageman, 2005) sheds light on how teams handle disequilibration or cognitive dissonance and emphasizes the importance of creating a safe environment for learning and adaptation within teams. Likewise, research on team conflict dynamics and effectiveness (O'Neill & McLarnon, 2018) notes the challenges of disequilibration that teams face.

8.3.2 Assimilation

Chris Argyris's concept of 'double-loop learning' (Argyris, 1993, 2004) uses a concept similar to metacognition in an individual, to explain how organizations assimilate new information and adjust their existing mental models to foster continuous improvement. Several authors who contributed to "The Fifth Discipline Fieldbook" (Senge et al., 1994) address how organizations assimilate new concepts such as systems thinking into their practices to enhance overall learning and adaptability.

8.3.3 Accommodation

Accommodation by the community (e.g. creating new knowledge and adjusting existing theories) can be triggered by Dewey's ideas. Impetus for change from experiential learning, for example, may involve adjusting the communal knowledge structures to accommodate and make sense of new empirical data. Collaboration and social interaction may require the community to adapt its norms and practices to accommodate additional diverse perspectives within its collaborative efforts. Inclusive learning may require adapting communal educational practices to accommodate the needs of a diverse learner population. The effect of these various kinds of accommodations causes a need for change in how the community understands its expertise, interacts with its members, assesses its structure of knowledge, and its processes for being able to know and act in its sphere.

Further support comes from Kurt Lewin's work in social psychology and group dynamics has implications for accommodation within groups. His model of unfreezing, changing, and refreezing (Lewin, 1946) suggests a process by which groups accommodate new information and practices. Edgar Schein's exploration of organizational culture and leadership (Schein, 2010) delves into how organizations accommodate new values, beliefs, and practices, leading to changes in the overall culture of the group. An entire industry has arisen to promote understanding of group dynamics influencing team learning and performance (Levi & Askay, 2021).

8.3.4 Equilibration

Recalling the need for both an individual and group interpretation at the meso level, there are two types of equilibrations: (1) The expert community achieves and then maintains a balance with other expert groups and plays a role in assuring its standards of knowledge and practice (e.g. how it knows and what it can do), and (2) individuals achieve and maintain membership in the social epistemology of the community (e.g. they become experts alongside others who share in the group's ways of knowing) while balancing their social position with their epistemological stance (e.g. their trajectory in the field and how they individually know and can do things an expert knows and does). In this sense Kolb's Experiential Learning Theory (Kolb & Fry, 1975) can be extended to teams and organizations, exploring how teams reach equilibration through cycles of concrete experience, reflective observation, abstract conceptualization, and active experimentation.

8.4 Implications of Dewey for Social Learning AI

Criteria for the development of AI applications consistent with and supportive of Dewey's ideas about social learning include environments, simulations, scenarios, tools, feedback, and themes focused on collaboration and other social contexts, and contain a balance of issues concerning the individual and the community (Table 8.1).

In summary, Dewey's pragmatic perspective on learning highlights the importance of social interaction, community engagement, and the incorporation of real-life experiences into education. He believed that learning was not only an individual cognitive process but also a social and communal endeavor that shaped individuals within the broader context of society. This dual purpose suggests a need for a theory that balances both individual and group concerns and raises the question of whether the causal mechanisms described for individual learning processes are valid for and have homologous functions in team and group learning. In the next chapter, cognitive science theory is harnessed to expand upon Dewey and create an encompassing computational framework for the social dimension of learning processes with links from the micro to the meso level as well as defining homologous functions of the PKK learning cycle for teams, organizations and other social learning settings.

8.5 Limitations and Conclusions

This chapter has not attempted to present an exhaustive analysis of social learning research. Instead, it has focused on the congruence of John Dewey's ideas with concepts in the PKK model of learning processes intending to assist in the process of applying both sets of ideas to a network model of the meso level. Dewey's

Table 8.1 Criteria for developing AI in education consistent with Dewey

Dewey's idea	Criteria	AI contribution
Experiential Learning	Design AI applications that prioritize hands-on, real-world experiences for users of value to some community.	Create virtual environments or simulations that allow users to actively engage and learn through experience. AI can provide realistic scenarios and feedback, enhancing the experiential aspects of learning like repeated practice, broadening experience to related problems and contexts, and receiving feedback on real-world performance.
Collaboration and Social Interaction	Foster collaboration and social interaction in learning environments.	Develop AI tools that facilitate collaborative learning experiences. Team projects and AI-supported group activities can encourage social interaction and shared learning experiences.
Individualized Learning Paths	Tailor education to the individual needs that balance with the interests of the community's knowledge and purpose.	Implement adaptive learning algorithms that analyze user behavior and preferences to personalize the approach of the community to sharing its content. AI can help create customized learning paths and adjust the difficulty level and content based on the learner's abilities and interests while balancing the needs of the community for developing expertise and expanding the community's knowledge base.
Integration of Formal and Informal Learning	Integrate formal education with real-life experiences.	Create AI applications that bridge the gap between formal and informal learning. This could involve incorporating real-world examples, simulations, or projects that connect academic concepts with practical applications and credentialling or validating learning from any setting.
Continuous Learning	Promote a lifelong learning mindset.	Develop AI-driven platforms that support continuous learning within the community. AI can recommend new topics based on a user or subgroup's interests, track progress, and encourage ongoing curiosity and exploration with others.
Critical Thinking and Problem-Solving	Emphasize critical thinking and problem-solving skills.	Implement AI-based scenarios and challenges that require users to think critically and solve problems with others. AI can provide feedback on the decision-making process of a group, encouraging iterative problem-solving, transparency, and equality of access to resources and opportunities to learn.
User-Centric Design	Prioritize the user's experience and needs viz mentoring and apprenticeship.	Use AI to analyze feedback and behavior from both newcomers and highly experienced experts to continually improve the application's usability for mentoring into and apprenticing within the community. Natural language processing, for example, can enhance communication between users and AI systems, making the learning experience more user-friendly.

(continued)

Table 8.1 (continued)

Dewey's idea	Criteria	AI contribution
Democratic and Inclusive Learning	Ensure access and inclusivity in learning opportunities.	Design AI applications that are accessible to diverse groups of learners and that promote ethical, inclusive, democratic processes. Use AI, for example, to provide multilingual support, accommodate different learning characteristics, and address individual as well as group learning needs concerning the roles and responsibilities of democracy and inclusive practices in education.

emphasis on experience, democracy, and social responsibility in education was shown to align with the Piaget-Keller-Kauffman (PKK) model's phases of learning. In addition, Dewey's pragmatic perspective on learning, which emphasizes the importance of social interaction, community engagement, and real-life experiences, interacts with the PKK learning cycle by highlighting the social and communal aspects of learning that shape individuals within the broader context of a social world.

The chapter suggested specific ways that Dewey's principles can be operationalized within AI applications to support collaborative learning environments. Additional readings from the literature on team learning and organizational behavior were aligned with the broad outlines of the PKK model, indicating the potential for the PKK mechanisms propelling learning at the individual level to play similar roles in team and organizational learning, which needs further discussion in the chapters that follow.

The PKK model's application to the meso level involves two types of equilibrations: one where the expert community balances with other expert groups and maintains standards of knowledge and practice, and another where individuals balance their membership in the social epistemology of the community with their epistemological stance. This dual interpretation at the meso level is consistent with Dewey's ideas, as he placed significant emphasis on the informal social context and dynamics of learning, considering them fundamental to the educational process.

Yet to be developed is the social extension of the aim articulated at the micro level, that the universal drive of learning is toward increased agency, self-awareness and self-regulation. What is socially shared agency of an expert group or learning community? What is team-based self-awareness and self-regulation? What are the social interpretations of the four PKK phases and mechanisms purported to drive universal learning processes?

8.6 Moving Forward

The social learning implications of an extension of the PKK model will be addressed in the next chapter along with a network model of the meso level. The meso level model has four nodes, including one node that contains multiple copies of the PKK model (e.g. one for each member of the team or organization).

References

Argyris, C. (1993). *Knowledge for action: A guide to overcoming barriers to organizational change*. Jossey-Bass Publishers.

Argyris, C. (2004). Double-loop learning and organizational change: Facilitating transformational change. In J. J. Boonstra (Ed.), *Dynamics of organizational change and learning* (1st ed., pp. 389–402). Wiley. https://doi.org/10.1002/9780470753408.ch19

Dewey, J. (1998). *Experience and education: The 60th anniversary edition*. Kappa Delta Pi.

Grand, J. A., Braun, M. T., Kuljanin, G., Kozlowski, S. W. J., & Chao, G. T. (2016). The dynamics of team cognition: A process-oriented theory of knowledge emergence in teams. *Journal of Applied Psychology, 101*(10), 1353–1385. https://doi.org/10.1037/apl0000136

Hackman, R., & Wageman, R. (2005). When and how team leaders matter—Hackman—Wageman—ROB 2004. *Research in Organizational Behavior, 26*(04), 37–74.

Kolb, D., & Fry, R. (1975). Towards an applied theory of experiential learning. In Cooper (Ed.), *Theories of group processes* (pp. 33–38). Wiley.

Levi, D., & Askay, D. A. (2021). *Group dynamics for teams* (6th ed.). SAGE.

Lewin, K. (1946). Action research and minority problems. *Journal of Social Issues, 2*(4), 34–46. https://doi.org/10.1111/j.1540-4560.1946.tb02295.x

O'Neill, T. A., & McLarnon, M. J. W. (2018). Optimizing team conflict dynamics for high performance teamwork. *Human Resource Management Review*. https://doi.org/10.1016/j.hrmr.2017.06.002

Schein, E. H. (2010). *Organizational culture and leadership* (4th ed.). Jossey-Bass.

Senge, P., Kleiner, A., Roberts, C., Ross, R., & Smith, B. (1994). *The fifth discipline fieldbook: Strategies and tools for building a learning organization*. Currency Doubleday.

Tuckman, B. W. (1965). Developmental sequence in small groups. *Psychological Bulletin*. https://doi.org/10.1037/h0022100

Tuckman, B. W., & Jensen, M. A. C. (1977). Stages of small-group development revisited. *Group & Organization Studies, 2*(4), 419–427. https://doi.org/10.1177/105960117700200404

Chapter 9
How People Learn

9.1 Introduction

The 4 node schema used for the meso level network was originally developed by John Bransford and colleagues and appeared in the early 2000s in the book *How People Learn* published by the National Research Council of the National Academy of Sciences (*How People Learn*, 2000). As a convention, the meso level will be identified as the HPL model.

The authors of the HPL model presented four issues—Knowledge, Community, Assessment, and Learner—as a comprehensive social cognition perspective on how people learn that will be used as the network nodes of the meso level (Fig. 9.1). Here and in Chaps. 10 and 11, we suggest that AI may play roles in each of the four dimensions. There are six bidirectional relationships in the graph such that each node communicates with the other three nodes via both inputs and outputs. If a modeling process needs to separate inputs from outputs, there are 24 relationships in the graph, six relationships for each of the four nodes.

The micro level PKK model of the individual is a subnet of the meso level called the *Learner* node and for modeling purposes, is replicated for each member of a team or organization. In the brief discussion of each node below, two perspectives will be advanced; first are the issues of importance to the team or organization and second are the issues of importance to a person seeking to become an expert in the community. These two stances expand upon the original model in Bransford and are carried forward into Chap. 10 to define goals for AI applications to enhance both individual and team-based learning.

© The Author(s), under exclusive license to Springer Nature Switzerland AG 2024

D. C. Gibson, D. Ifenthaler, *Computational Learning Theories*, Advances in Analytics for Learning and Teaching,
https://doi.org/10.1007/978-3-031-65898-3_9

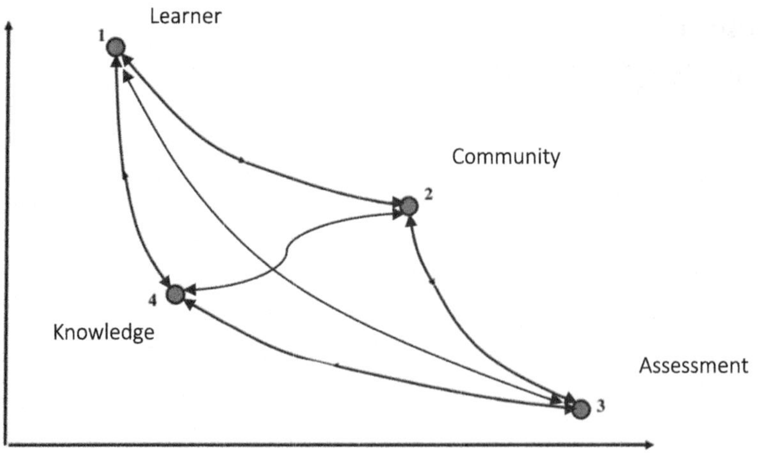

Fig. 9.1 Meso level network model

9.1.1 Knowledge Is Dynamic and Collaboratively Determined

Knowledge held by a community of experts in any field includes an emphasis on the collaborative and social nature of knowledge construction (e.g. what counts as legitimate knowledge and how it is discovered, vetted, validated and adapted) as well as its relatively static collection of facts. Bransford and his cognitive science author colleagues highlight that knowledge is not just an individual possession but is actively constructed and negotiated within a community of learners. In a knowledge community, learners are not passive recipients of information but active participants in the co-construction of knowledge within a supportive and interactive learning environment. The social and contextual aspects of meso level learning thus suggests a need for induction and instructional practices that leverage the dynamic and collaborative nature of knowledge development within authentic communities of practice.

Consistent with the micro level PKK model, according to Bransford, individuals in teams or organizations build a private world of knowledge on existing mental models and prior knowledge. Learners organize and relate new information to their existing understanding and for learning to be meaningful, content must be relevant, connected, and aligned with students' prior experiences. Importantly for social relevance and validation, learners must be engaged in activities that mirror the practices (e.g. the knowledge-in-action) of experts within a particular field. This involves immersing learners, for example, in *cognitive apprenticeships* (Resnick, 1984) of processes of inquiry, problem-solving, and collaborative discourse that experts use to build and refine their knowledge. Bransford underscores the idea that learning is not just about acquiring isolated facts but, as Dewey advocated, includes understanding how knowledge is interconnected and applicable in real-world contexts.

9.1.2 Community Experts Own the Knowledge and Learning Processes

The community perspective underscores the idea that meso level learning is not solely an individual endeavor but emerges from interactions within an interactive knowledge community situated in an authentic enterprise of applying its knowledge in a real-world setting. Novices benefit from participating in a community where they can observe, interact with, and receive guidance from experts, where they develop situated cognition of the knowledge and practices of the community (Brown et al., 1989). It has been suggested that generative AI can play the role of a community expert, for example, as a conversational agent participating in educational dialogs (Sharples, 2023). As the community inducts new members, it must provide mentoring through authentic tasks, case studies, and collaborative activities that mirror the ways its experts think and work (Robbins, 1999). The goal is to help learners transition from novices to more advanced levels of understanding by actively participating in discussions, sharing perspectives, and collaborating on projects. A primary responsibility of a knowledge community is then creating learning environments and opportunities that reflect the authentic practices of its discipline.

From the perspective of the individual, a knowledge community is a group of experts who have knowledge and skills that one might wish to obtain. Importantly, expertise is not just about possessing a large store of facts; it also involves a deep understanding of the underlying principles and patterns within a domain (Ericsson et al., 2006; Sternberg, 1999). So, experts in the community, who have a robust mental framework that allows them to efficiently organize and apply their knowledge, need to expose newer individuals to these expert-like practices, and newer learners need to be encouraged and supported to think like and learn from their actions as emerging professionals in the field.

9.1.3 Assessment Feedback Develops Community Expertise

Assessment practices play a crucial role in improving the knowledge community by providing valuable insights into the understanding, performance, and development of individuals within the community vis a vis the community's knowledge and expertise. Assessment can identify gaps in the understanding of individuals, provide data for refining teaching strategies and materials, and foster a culture of ongoing improvement. Assessment practices involving self-assessment and reflection encourage all community members toward improvement of the community's knowledge and learning processes. This *collective metacognitive awareness* is valuable for fostering a deeper understanding of each member's strengths and weaknesses, promoting autonomy, and cultivating a sense of responsibility for personal and collective growth. Collaborative assessments and peer feedback mechanisms also promote learning within the community. Experts can use peer assessments to encourage interaction, dialogue, and the sharing of diverse perspectives, which advances knowledge, collective expertise and a sense of community. In summary, assessment practices serve as a dynamic tool for enhancing the community of experts by offering insights into both individual and collective learning processes, guiding instructional improvements, and fostering a culture of continuous reflection and growth within the community (Black & Wiliam, 2009).

The individual's stance toward assessment, in contrast, is often viewed primarily as a tool for grading and comparing one's expertise status within the community but should also be a significant and integral part of the learning process. Bransford thus advocates for formative assessment, which provides ongoing feedback to both the community of experts (e.g. what is working, how new members are faring) and learners (e.g. the extent to which members are becoming experts), informing instructional decisions and promoting continuous improvement. Assessment, in this context, is synonymous with effective critical feedback and becomes a means to understand and support the learning journey of the individual within the community rather than just a final judgment (Shute, 2008).

9.1.4 Learners Are the Creative Force of Knowledge Communities

The community perspective on individual learners focuses on expanding the society's numbers as well as its capabilities, because creative extensions of knowledge can, and often do, come from new members as well as the current experts (Csikszentmihalyi, 1996). The current members of the community might do more to hold on to old paradigms, even as mounting evidence of a need for new ideas surface and a paradigm shift is brought to the attention of the experts (Kuhn, 1962). So, induction and mentoring of new members into the community is crucial for ongoing renewal of ideas and improvements in practices, yet the activity must be balanced with the guardianship activities of quality control, ensuring theoretical cohesiveness, and promoting the practical real-world effectiveness of the community.

The learner-centered approach is a central theme in Bransford, which recognizes the diversity of learners and their unique needs. At the heart of individual agency and self-regulation of learning, Bransford focuses on metacognition, the capabilities and habits of students reflecting on their thinking processes and developing strategies for effective learning. Bransford also addresses the role of motivation and interest in driving successful learning outcomes, so common ground with the PKK model can be found in abundance. The PKK model provides a subnetwork model of the internal details about the mechanisms of learning that underlie metacognition, which fuels the growth of agency, self-awareness and self-regulation. At the meso level, the individual learner is a group member participating (and sometimes initiating) in additional roles such as generating alternative ways of expressing, responding to an argument with either support or criticism, making suggestions that advance collaboration, co-designing thought (or real) experiments, exploring and analyzing data, and creating narratives. These roles can be played by large language model AI (LLMs) underscoring the social role of generative AI at the meso level (Sharples, 2023).

The four HPL nodes of the meso level (i.e. knowledge, community, assessment and learner) constitute a holistic understanding of social learning processes that balance individual and collective aspects of learning processes in social settings like teams and organizations. In Chap. 10 the four nodes of the meso level network will be used as a structure to imagine AI applications that can be sensitive to and take actions to assist individuals (or AI acting as an individual) and in Chap. 11, to assist the team or knowledge community in its roles. AI in team learning is primarily constrained by each field's knowledge and practices, but can act creatively to increase the agency, reflective awareness and regulatory functions of both the community and its members, while leveraging the individual learning cycle for maximum effectiveness of the learner within the learning community. Before dealing with these implications of the meso level model, the narrative addresses the question of how the learning cycle changes or needs reinterpretation when the objective of increasing agency changes from an individual to a group, and how the fully connected network balances the needs of individual members with the collective aims of the practitioners of a field of knowledge.

9.2 Homologous Functions of the Learning Cycle at the Meso Level

To support the social extension of the aim articulated at the micro level—that the universal drive of learning is toward increased agency, self-awareness and self-regulation—this section discusses the homologous functions of the PKK model at the meso level. The discussion aims to address questions like: What is the socially shared agency of an expert group or learning community? What is team-based self-awareness and self-regulation? What are the social interpretations of the four phases and mechanisms purported to drive universal learning processes?

To review the PKK learning cycle functions:

1. Disequilibration phase leads to Attention—an external (isothermal expansion) event triggers arousal of noticeable differences.
2. Assimilation phase leads to Relevance—an internal information search (adiabatic expansion) expands the internal representation space and linkages and identifies meaningful gaps.
3. Accommodation phase leads to confidence—new symbols and relationships are created (isothermal compression) to fill the gaps and when externalized, the environment then provides feedback concerning the updated internal models (e.g. knowledge, skills, know-how, expertise).

4. Equilibration phase leads to satisfaction—internal attractors (e.g. set points, cycles and assemblies) are reset (adiabatic compression) to a new normal of monitoring via ongoing cycles of prediction and validation.

In the PKK model, the relationships linking the nodes was directional (i.e. the network graph was a directed cyclic graph) to represent the causes in the cycle, but at the meso level, the links become contingent. The influence direction is now bidirectional when considering the network (more complex if including influences from the PKK model) and the causes of a state of the network may be difficult to ascertain except in a specific context or trajectory within the network.

For example, the node labeled *knowledge* is often thought of as static but must change from time to time as new facts arise and theories are combined, extended, or abandoned (Kuhn, 1962). The inputs to knowledge in the meso level model come from either the community or assessment, not directly from learners except as their contributions are vetted via the community or an assessment. Partial causality therefore can be ascribed at every node from every other node, and it remains to be tested if the probable influences from either direction have predictable differences in magnitudes (e.g. stronger correlations, or orders of principle component analyses) related to their distance from the effect. For example, if the community's *knowledge* needs to be updated, which came first: the community, then a specific learner or subgroup, then an assessment? or was the need for new knowledge first encountered due to an assessment, impacting a member or subgroup, and eventually the community as a whole? Appropriate multifactorial empirical studies may establish different paths for each specific context. In later chapters, the bi-directionality of the meso-level links will be expanded to be fully multidirectional as well as multimodal, increasing the possibilities for influences across the entire network of networks at the macro level.

Knowledge at the meso level is the socially shared agency of an expert group or learning community and refers to know-how as well as know-what, the epistemic experiences as well as the epistemology of the domain of the experts. In the micro level, the homologous node was the accommodated mental model being maintained during dynamic equilibration. So expert knowledge at the meso level is likewise dynamic yet stabilized from the history of the community's experience of how it has come to know and continues to know its subject and processes.

Disequilibrations that shake the foundations of knowledge normally do not come from an individual who is not a full member of the community, as creativity itself and all innovative ideas are vetted and understood as belonging to and being relevant to the existing shared mental model of the community (Csikszentmihalyi, 1996). Thus, the community is a mediator of advances in knowledge. At the same time, a diversity of individuals make up the community, and some of these people might be recent arrivals, bringing with them new knowledge and skills with novel vocabulary and ways of knowing, carrying with them new connections from fields not yet connected to the community.

Assessment of innovative individual contributions, whether by highly experienced and politically connected experts or new arrivals, provides a mechanism for testing and vetting ideas. Suppose that team-based self-awareness and self-regulation are roles of community assessment practices.

These kinds of considerations concern the social interpretations of the four phases and their mechanisms which the theory asserts drive universal learning processes. Knowledge is maintained in dynamic equilibrium with the progress of other fields and the explorations and assessments by experts in the community. The expert community is the context for moments of disequilibrium, as any challenge to theory or new empirical evidence may upset the status quo. Assessment practices involve some members in helping others to assimilate and other members who are at some stage of assimilating into the community. Finally individual members of the community are the motive force of accommodation by the community because the entire edifice rests on their shared understanding, processes and assessment of evidence.

This brief overview of the homologous functions of the learning cycle introduces an additional complication. For each member of the team or community, there are at any moment, influences from their internal learning journey(s) interacting with other people in the community. While these interactions are not uniquely causal at the meso level compared to the individual learning process, the expressed meso level interactions of each member (e.g. with other members, the knowledge base, as well as assessment and community practices) can contribute to setting the conditions for more probable next stages in the social dynamics.

For example, there is a causal relationship between effective feedback mechanisms and continuous improvement. Constructive feedback contributes to learning and improvement within a team, driving a cycle of feedback leading to enhanced performance. This implies that if one or more members of a team are effective in giving feedback, the team's performance will be better than otherwise. Similarly, psychological safety, a positive characteristic of certain community norms, can contribute to open communication within a team. When team members feel psychologically safe, they are more likely to be transparent and honest in their communications. There is also an influential relationship between expertise and guidance within a team. When experts provide effective guidance, a feature of the assessment node, to less experienced members, it facilitates knowledge transfer and skill development. Likewise, diversity of individual talents and perspectives within a team or community can influence innovation because exposure to different viewpoints and backgrounds stimulates creativity and can lead to novel problem-solving approaches.

The concept of cognitive apprenticeship (Collins et al., 1989; Resnick, 1984), another potential characteristic of a knowledge community, implies a causal relationship between learning and skill transfer from more experienced individuals. Less experienced team members can acquire skills through observation, imitation, and collaboration with experienced counterparts. Shared mental models like the knowledge content of the experts of the community causally contribute to improved coordination within a team but might pose a barrier to new ideas and evolution of the field. A common understanding of goals, processes, and strategies enhances communication and ensures that team members are on the same page, but uncritical acceptance of orthodox positions can close the community to innovations in processes as well as new understandings.

It's important to reiterate that the relationships at the meso level among knowledge, community, learners and assessment, are not one-directional, as were the relationships in the directed cyclic graph of the individual learning cycle. At the meso level there are reciprocal influences among the elements, complicating the issue of causal precedence; for example, to what extent does knowledge shape the community or does the community shape its knowledge? The answer may change from context to context. Additionally, the effectiveness of these relationships may be influenced by not only contextual factors, but also team dynamics, and the specific goals of the team.

9.3 Limitations and Conclusions

The HPL or meso level model's dynamics were not explained in detail, as there are a very large number of possible paths for analysis. As each node has 3 inputs and 3 outputs and the relationships are bidirectional, a four-step process could have any of 81 possible trajectories starting from any node. That is, starting from say a learner, the trajectory can go to any of the three remaining nodes, which is repeated at each step of the four-step process. To narrate the possibilities would be onerous and there is no rationale for why the process should be four or some other number of steps.

The section discussing the homologous functions made an informal case for a 'most probable' loop of the cycle IF the meso level in a particular situation had a dynamic like the causal learning cycle of an individual. For example, if a new *learner* introduced an idea unlike anything the community had seen before, it could *disequilibrate* the community's ongoing *equilibration* that maintains its grasp of *knowledge,* but only if the community *assesses* the new idea as both *relevant* and requiring both *assimilation* and *accommodation*. This informal case, which seems coherent and consistent with the micro level causal model, is not determinative because 80 other equally coherent and consistent stories could be told. So, one may ask of what value is the meso level model?

One answer to this question is contained in the next chapter, where the model's elements and features are used to guide the development of AI to intervene or collaborate with a community to increase the community's agency (e.g. power to make decisions and act on its behalf in the world), collective awareness and self-regulatory capabilities. Another possibility is that by limiting the network to four nodes, it might facilitate the compression of ideas and discovery of commonalities across fields that are tangent to or interwoven with team and organizational learning, which may lead to new understandings and cross-disciplinary findings and bring order to one or more fields.

A downside of having only four nodes is that each one needs further delineation of internal structures and relationships, which is true even for the node labeled '*learner*' which already has a well-defined internal structure and dynamic relationships but needs to be studied further in the case where two or more learners interact with each other. For example, what are the impacts on one individual's internal states viz. an influence arrives from another person, and when do the causal forces of one accelerate or decelerate those of the other? There are undoubtedly clues, evidence and opinions about many of the possibilities in existing literature, which the micro and meso level models could help organize.

Despite these limitations, there is now a two-thirds completed computationally friendly model that is consistent with extant scientific research conclusions, in the form of completed descriptions of a micro and meso level. The micro level network model is a subnet of the meso level model. The levels can provide context for each other (e.g. variable representing the nodes and algorithms representing the processes can either include or ignore either level in an analysis of the other) or be treated as separate isolated descriptions for modeling purposes.

9.4 Moving Forward

This chapter presented the network model of the meso level or HPL as containing four nodes in a fully connected bidirectional network of 24 relationships. The network model was developed consistent with Dewey and Bransford's learning theories as beacons of the cognitive science model of the social construction of knowledge, which is discovered, created, shared and evolved via communities of practice. Those social learning processes rest on and in a sense scale individual learning processes. The relationships in the meso level model are probabilistic and partially causal due to the bidirectionality of influences in the network. In the next chapter, the implications of the HPL network model help develop a set of specific idea for the development of AI to enhance learning processes of both the individuals and groups involved at the meso level.

References

Black, P., & Wiliam, D. (2009). Developing the theory of formative assessment. *Assessment Evaluation and Accountability, 1*(1), 5–31. https://doi.org/10.1007/s11092-008-9068-5

Brown, J., Collins, A., & Duguid, P. (1989). Situated cognition and the culture of learning. *Educational Researcher, 18*(1), 32–42. https://doi.org/10.3102/0013189X018001032

Collins, A., Brown, J., & Newman, S. (1989). Cognitive apprenticeship: Teaching the crafts of reading, writing, and mathematics. In *Knowing, learning, and instruction: Essays in honor of Robert Glaser* (pp. 363–401). MacMillian.

Csikszentmihalyi, M. (1996). *Creativity: Flow and the psychology of discovery and invention.* HarperCollins.

Ericsson, K. A., Charness, N., Hoffman, R. R., & Feltovich, P. (2006). *The Cambridge handbook on expertise and expert performance*. Cambridge University Press.

How People Learn (2000) *Brain, mind, experience, and school* (Expanded ed., p. 9853). National Academies Press. https://doi.org/10.17226/9853

Kuhn, T. (1962). *The structure of scientific revolutions*. University of Chicago Press.

Resnick, L. B. (1984). Education and learning to think. In *Learning*. Falmer Press.

Robbins, P. (1999). Mentoring. *Journal of Staff Development, 20*(3), 40–42.
Sharples, M. (2023). *Towards social generative AI for education: Theory, practices and ethics.* https://doi.org/10.48550/ARXIV.2306.10063
Shute, V. J. (2008). Focus on formative feedback. *Review of Educational Research, 78*, 153–189. https://doi.org/10.3102/0034654307313795
Sternberg, R. J. (1999). Intelligence as developing expertise. *Contemporary Educational Psychology, 24*(4). https://doi.org/10.1006/ceps.1998.0998

Chapter 10
AI Assisting Individuals as Team Members

10.1 Introduction

At the meso level, the roles of AI are complex because the roles need to balance (1) tracking and responding to everyone simultaneously within a focus of collaboration during team problem-solving and knowledge building, and (2) assisting the team in working within (or breaking tradition with) a larger community of inquiry or field of research knowledge. If the former does not happen, then the team or the teamwork is more likely to fail, and if the latter does not happen, then the team's work will not be accepted by or related to a community of inquiry or authority of experts. In this chapter the focus is on the first task of assisting individuals as team members and partnering with humans on their goals and tasks.

When an individual is a member of a team, the focus of learning shifts from acquiring and practicing knowledge and abilities to fulfilling social roles in collaboration. As pointed out by others, AI systems need to be able to understand the learners' goals and values, respect their autonomy, and provide appropriate and timely support (Sharples, 2023). AI can assist the person with these roles in several ways aligned to the four nodes of the meso level: learner, community, assessment, and knowledge (Table 10.1).

10.1.1 Learner: Matching Individual Capabilities to Task Requirements

Human-AI teams will be more effective if team or organization members can understand the behaviors of the AI system, develop trust relationships with the AI, make accurate decisions based on input from the AI and exert timely control over the

© The Author(s), under exclusive license to Springer Nature Switzerland AG 2024
D. C. Gibson, D. Ifenthaler, *Computational Learning Theories*, Advances in Analytics for Learning and Teaching,
https://doi.org/10.1007/978-3-031-65898-3_10

Table 10.1 AI assisting individuals in collaborating with others and partnering as a team member

HPL Model of the meso level	AI assisting individuals as team members and joining the team as a member
Learner—social status, demographics, characteristic personality, emotional, behavioral, cognitive components, intentions *cause or mediate* **Disequilibration** (or goes unnoticed by the community)	AI matches individual capabilities to task or group goal requirements and adds unique individual member value (e.g. computation, data analysis, natural language cognition) to the team.
Community—the gatekeepers of the fields of knowledge, the community of practice *causes or mediates* **Assimilation**	AI nudges each team member to fulfill collaboration roles and fulfills roles like coordinator, orchestrator, or human-resource manager.
Assessment—feedback from novice to expert, admission to the field, determines individual and group fitness and importance to the field *causes or mediates* **Accommodation** (if the learner is accepted along with their contributions)	AI assesses and assists members in maintaining cognitive presence and continuous learning and nudges the team to maintain focus on goals and tasks.
Knowledge—socially constructed, models, hypotheses, empirical, symbolic, semantic. The epistemic experience of field experts builds this knowledge. *Causes or mediates* **Equilibration** (if the field of knowledge adapts to or ignores the learner's contributions)	AI assists members in fulfilling problem-solving and knowledge-building roles and contributes to the co-creation of knowledge.

system (NAS et al., 2022). Two areas where AI can make a significant impact are helping to form teams with complementary skillsets and knowledge and assisting the team to effectively utilize those strengths and differences to address problems and challenges.

10.1.1.1 Skill Mapping

AI can help analyze individual strengths, skills, and preferences, aligning individual team members with task demands or project needs within the community. These roles involve leveraging data analytics to comprehensively analyze individual strengths, skills, and preferences within the team or community and then matching those with task characteristics via a recommender system. For example, machine learning algorithms can process diverse datasets, including resumes, project histories, and performance metrics, to create a detailed skill profile for each team member. An approach that combines transformer architecture influenced by recent large language models with natural language content awareness (Lin et al., 2022) suggests that future team building systems could match team skillset combinations with problem characteristics helping to optimize both team formation and project outcomes.

10.1.1.2 Task-Skill Allocation

Once teams are formed around a problem or challenge, AI can aide in task allocation by utilizing individual skill profiles to recommend tasks or roles that align with each team member's capabilities. In allocating between humans and AI, machine learning models can analyze historical data, performance metrics, and feedback to match capabilities with needs at both general and task-specific levels (Parasuraman et al., 2000).

10.1.2 Community: Fulfilling Collaboration Roles

Collaboration roles include helping the team establish and maintain shared understanding, taking appropriate action to solve the problem, and establishing and maintaining team organization (Roschelle & Teasley, 1995) and future AI may fill those roles as well as flexibly alter its role to boost team performance by serving as a team coordinator, orchestrator, or human-resource manager (NAS et al., 2022).

10.1.2.1 Collaboration Facilitation

AI can enhance collaboration in teams by leveraging data-driven insights to identify and capitalize on complementary skills among team members. For instance, machine learning algorithms can analyze individuals' expertise, experiences, and working styles, helping to form sub-teams and task teams that possess a diverse set of skills. In addition, interaction dynamics, such as turn-taking, contribution type, and communication, which are among the driving forces of the co-creative process can be included in the design of AI applications (Rezwana & Maher, 2023).

In a recent research project, a measure of network influence that captured skill transfer over time among team members was used to design a recommendation system to suggest new teammates based on a modified deep neural autoencoder (Sapienza et al., 2019). This is evidence that AI tools can potentially foster connections among team members by recommending potential collaborators based on shared interests, past collaborations, or complementary skills. As AI becomes more capable and autonomous, it can become a teammate in certain situations (NAS et al., 2022).

10.1.2.2 Task-Specific Collaboration Support

AI can offer real-time collaboration guidance and facilitate collective problem-solving via task allocation. Natural language processing algorithms, for example, can analyze team conversations, project-related communications, and individual contributions and then offer insights on communication patterns and potential areas of improvement. Collaborative task allocation in human-AI teams also sheds light on the potential of AI in optimizing creativity and problem-solving within diverse groups and personalized task allocation has been found to enhance team efficiency and individual fulfillment (Ali et al., 2022). In elaborating roles for AI teammates a recent study used exploratory and confirmatory factor analysis of expert opinion concerning team roles to construct sets of behaviors and skills that AI-based teammates can fulfill in teams: the coordinator, creator, perfectionist and doer (Siemon, 2022).

As an example of the collaborative intelligence shared by AI and humans on teams, a fine-grained human digital twin modelling method has been introduced as well as a demonstration-based learning approach to handle task uncertainties (Zheng et al., 2023). So, it seems likely that future AI applications will participate in crowdsourcing approaches to task-matching systems based on implicit profiles and performance history of users. Beyond the initial allocation of tasks, AI can also challenge and engage team members appropriately during learning processes. This personalized approach fosters a sense of accomplishment and satisfaction, contributing to higher levels of engagement and success within the team.

Moreover, AI-powered tools can provide task-specific resource recommendations, such as suggesting relevant documents or experts, to enhance the team's problem-solving capabilities. By offering real-time support tailored to the task at hand, AI contributes to smoother interactions and more effective collective problem-solving within teams. Team-centered AI needs to align itself to the team or organization's goals and then participate in the communication and decision making with humans. Recent research has begun to delineate the teamwork requirements for successful human-AI teaming in interdependent and complex work domains, which includes responsiveness, situation awareness, and flexible decision-making. Technical requirements have been also noted, including cognitive competence, reinforcement learning, and semantic communication (Hagemann et al., 2023).

10.1.3 Assessment: Maintaining Cognitive Presence and Team Focus

An ongoing assessment that helps learning teams stay focused on their work could be described as a process of monitoring cognitive presence (Garrison et al., 2009) which before ubiquitous AI, was determined by a post-hoc analysis of text to classify the nature and quality of critical discourse and thinking. Classification of natural language is now relatively easy for a large language model. Researchers have noted that appropriate cognitive presence (e.g. exploring, integrating, resolving) and social presence (e.g. exhibiting affective expression and open communication) helps create supportive conditions (Khot et al., 2022). Assessing and nudging team members on presence and focus is now also feasible for AI, as demonstrated by Ashol Goel and colleagues at Georgia Tech's creation of "Jill Watson"—an online course assistant (Wang et al., 2020). AI can also utilize digital signatures aligned to empirically validated instruments to assess discourse, actions, and relationships to both others and tasks to take actions to help the team or an organization. These perspectives, focused on continuous assessment and feedback, rather than conceptions of AI in assessment concerning accountability or cheating (González-Calatayud et al., 2021), can help maintain a focus on team-based learning processes.

10.1.3.1 Cognitive Monitoring

AI can observe and interpret engagement patterns, intervening when individuals show signs of disengagement or cognitive overload while conducting a continuous assessment of a team's progress and helping to ensure shared understanding and commitment among group members. For instance, machine learning algorithms can analyze team interactions, project-related communications, and collaborative activities to provide real-time insights into the team's dynamics. A recent study exemplified this when it analyzed automated group learning engagement via deep neural network models in a computer-supported collaborative learning context. The research found significant differences in experimental and control group learning engagement, group performance, collaborative knowledge building, and socially shared regulation between experimental and control groups (Zheng et al., 2023).

10.1.3.2 Adaptive Support Systems

AI-driven assessments can not only provide valuable feedback on team dynamics and progress but can also offer recommendations to enhance collaboration and maintain a shared understanding of the problem at hand. Through continuous monitoring and analysis, AI can contribute to fostering a more cohesive and effective team environment. AI can provide prompts, resources, or breaks based on cognitive load assessment, ensuring sustained engagement and focus. Additionally, AI-powered tools can assist in maintaining focus on problem-solving by monitoring the team's progress and suggesting adjustments to strategies based on real-time data.

10.1.4 Knowledge: Fulfilling Problem-Solving and Knowledge-Building Roles

Knowledge-building by a team involves stages like exploring a challenge or problem, reaching a common understanding with others about the nature of the problem and its features, representing and formulating the problem, planning and executing an approach that addresses the problem, monitoring progress toward the solution or effects of the plan, and reflecting on the results (Mayer & Wittrock, 1996). Two practical roles for AI that would assist a team or organization are assisting with access to knowledge repositories and generating ideas during the stages of ideation.

10.1.4.1 Knowledge Repository Access

Smart applications can offer access to relevant knowledge repositories, aiding in team problem-solving and supporting continuous learning by the team. Generative AI can automate content generation and create syntheses of information. For instance, language models like OpenAI's GPT-4 can assist in the creation of reports, summaries, and even educational materials by generating coherent and contextually relevant text. This capability is highlighted in Brown et al.'s paper "Language Models are Few-Shot Learners," showcasing GPT's ability to perform various language tasks with minimal examples (Brown et al., 2020). Moreover, AI can streamline the process of extracting valuable insights from vast knowledge repositories. Natural language processing algorithms can be employed to analyze and summarize large volumes of text, allowing teams to quickly identify relevant information. This is echoed in the work of Mikolov et al. in "Distributed Representations of Words and Phrases and their Compositionality," demonstrating the potential of word embeddings for semantic analysis (Mikolov et al., 2013). AI can thus not only enhance information retrieval but also facilitate efficient knowledge utilization by teams.

10.1.4.2 AI-Supported Ideation

Generative AI can foster creativity and knowledge-sharing within teams. For instance, large language models can assist in brainstorming sessions by generating a diverse range of ideas across different domains. In idea and knowledge creation, AI can use predictive analytics and conduct self-learning analysis, recognize previously unknown patterns as well as classify examples, sift through organizational data to discover relationships, and even develop new ideas for declarative knowledge using generative large language modeling (Jarrahi et al., 2023). In addition, teams can benefit from AI's assistance in storage and retrieval of ideas, sharing results, and generating ideas for applying knowledge. Furthermore, AI can facilitate knowledge-building by aiding in the creation of educational content, collaborative documentation, and even tutorial materials. By leveraging generative models, teams can automate the generation of coherent and informative content, accelerating the knowledge-sharing process. In essence, AI-generated ideas can contribute to brainstorming sessions (Habib et al., 2024), while knowledge-sharing benefits from AI as a partner can result in efficient creation and dissemination of information within collaborative environments.

10.2 Limitations and Conclusions

The focus of this chapter has been on AI playing a role as a team member, so has been limited to an analysis of individual roles in collaboration and problem-solving. In the next chapter the focus will be on AI working on behalf of the organizational owner of the team or group. This shift from individual to collective goals mirrors the shift that takes place from the micro (individual) to the meso (social) level. The chapter did not extensively discuss the potential challenges or limitations of integrating AI into team settings, such as issues related to privacy, ethics, or biases that may arise when using AI in decision-making processes. It also did not provide a comprehensive analysis of the potential drawbacks or risks associated with relying heavily on AI in team collaboration, such as over-reliance on AI recommendations or the potential for reduced human agency and decision-making autonomy.

As the focus was on the individual learning processes of team or group membership, the chapter did not explore the potential limitations or challenges in implementing AI systems within organizational contexts, such as the need for infrastructure, training, and ongoing maintenance and support. Nor did it address the potential impact of AI on job displacement or changes in team dynamics and social interactions within teams. These topics are more relevant in Chap. 11 as the focus shifts from the individual to the collective.

The chapter has highlighted the importance of individual team members understanding the behaviors and limitations of AI, developing trust relationships by limiting AI's role, making accurate and informed decisions, and exerting timely control over the AI as a team member. AI was discussed in terms of effectively utilizing the strengths and differences of members to address problems and challenges. The chapter also discussed how AI can enhance collaboration by not only leveraging data-driven insights to identify and capitalize on complementary skills among team members but also using that information to form sub-teams and task teams with diverse skill sets. The chapter also highlighted the role of AI in assisting individuals with fulfilling social roles in collaboration within a team, shifting the focus of learning to acquiring and applying knowledge in a social setting. The various perspectives on AI in this chapter's scope were shaped by a four-part framework advanced by John Bransford and colleagues that includes individuals who aspire to become experts in a domain, a community of people that develops expert knowledge and practices, the community's ability to assess and give feedback concerning an individual's level of expertise, and domain knowledge (e.g. learner, community, assessment and knowledge).

10.3 Moving Forward

At the meso or social level of learning processes, AI has a dual role as an assistant to individuals and an assistant to an expert community. The influences in the network model are less constrained, more fully connected and ambivalent at the meso level compared to the micro level. So, modeling with network analytics is flexible but contains more uncertainties than the simpler causal model of the micro level. As a fully connected graph of four nodes, with bivalent connectors, the number of trajectories climbs rapidly as the number of steps in a process increases.

For example, in a three-step process that includes all four nodes (e.g. which is the minimum process for including all four nodes), there are 4*3*2 = 24 trajectories. These 24 may be what Senge and others have referred to as *system archetypes* (Kim, 1994), which will be discussed in Chap. 11. Adding a fourth step results in 4*4*4*4 = 256 trajectories, since the assumption of involving all four nodes means that at every step, a node can influence itself (e.g. via reentry signaling, or reflection). Longer processes with more steps exponentially increases the adjacent possibility space, so we must abandon the project of discussing each case and instead turn toward principles that hold promise for understanding the dynamics and that guide an intuitive sense of the learning processes. This is the project of Chap. 11, to suggest AI roles in learning processes that assist the expert community as a collective enterprise.

Of interest in Chap. 11 concerning the collective level is how AI can work on behalf of the team or group to achieve collective goals and how to better understand the potential changes in individual roles and responsibilities within teams when AI is a team member. More research is needed on how AI providing real-time insights into team dynamics, project-related communications, and collaborative activities enables or impedes teams in optimizing their performance and effectiveness. Chapter 11 aims to discuss these issues and to suggest ideas for the development of AI tools and algorithms that can enhance knowledge sharing, assessment, and community building within teams.

References

Ali, A., Tilbury, D. M., & Robert, L. P. (2022). *Considerations for task allocation in human-robot teams*. https://doi.org/10.48550/ARXIV.2210.03259

Brown, T., Mann, B., Ryder, N., Subbiah, M., Kaplan, J., Dhariwal, P., Neelakantan, A., Shyam, P., Sastry, G., Askell, A., Agarwal, S., Herbert-Voss, A., Krueger, G., Henighan, T., Child, R., Ramesh, A., Ziegler, D. M., Wu, J., Winter, C., et al. (2020). *Language models are few-shot learners*. https://doi.org/10.48550/ARXIV.2005.14165

Garrison, D. R., Anderson, T., & Archer, W. (2009). Critical thinking, cognitive presence, and computer conferencing in distance education. *American Journal of Distance Education, 21*(1), 7–23. https://doi.org/10.1080/08923640109527071

González-Calatayud, V., Prendes-Espinosa, P., & Roig-Vila, R. (2021). Artificial intelligence for student assessment: A systematic review. *Applied Sciences, 11*(12), 5467. https://doi.org/10.3390/app11125467

Habib, S., Vogel, T., Anli, X., & Thorne, E. (2024). How does generative artificial intelligence impact student creativity? *Journal of Creativity, 34*(1), 100072. https://doi.org/10.1016/j.yjoc.2023.100072

Hagemann, V., Rieth, M., Suresh, A., & Kirchner, F. (2023). Human-AI teams—Challenges for a team-centered AI at work. *Frontiers in Artificial Intelligence, 6*, 1252897. https://doi.org/10.3389/frai.2023.1252897

Jarrahi, M. H., Askay, D., Eshraghi, A., & Smith, P. (2023). Artificial intelligence and knowledge management: A partnership between human and AI. *Business Horizons, 66*(1), 87–99. https://doi.org/10.1016/j.bushor.2022.03.002

Khot, N., Sarkar, M., Bansal, U., Singh, J. V., Pharande, P., Malhotra, A., & Kumar, A. (2022). Community of Inquiry framework to evaluate an online obstetric and neonatal emergency simulation workshop for health professional students in India. *Advances in Simulation, 7*(1), 25. https://doi.org/10.1186/s41077-022-00220-5

Kim, D. H. (1994). *Systems archetypes*. Pegasus Communications.

Lin, W., Shou, L., Gong, M., Jian, P., Wang, Z., Byrne, B., & Jiang, D. (2022). *Transformer-empowered content-aware collaborative filtering*. https://doi.org/10.48550/ARXIV.2204.00849

Mayer, R., & Wittrock, M. (1996). Problem-solving transfer. In D. Berliner & R. Calfee (Eds.), *Handbook of educational psychology* (pp. 47–62). Simon & Schuster Macmillan.

Mikolov, T., Sutskever, I., Chen, K., Corrado, G., & Dean, J. (2013). *Distributed representations of words and phrases and their compositionality*. https://doi.org/10.48550/ARXIV.1310.4546

NAS, Committee on Human-System Integration Research Topics for the 711th Human Performance Wing of the Air Force Research Laboratory, Board on Human-Systems Integration, Division of Behavioral and Social Sciences and Education, & National Academies of Sciences, Engineering, and Medicine. (2022). *Human-AI teaming: State-of-the-art and research needs* (p. 26355). National Academies Press. https://doi.org/10.17226/26355

Parasuraman, R., Sheridan, T. B., & Wickens, C. D. (2000). A model for types and levels of human interaction with automation. *IEEE Transactions on Systems, Man, and Cybernetics – Part A: Systems and Humans, 30*(3), 286–297. https://doi.org/10.1109/3468.844354

Rezwana, J., & Maher, M. L. (2023). Designing creative AI partners with COFI: A framework for modeling interaction in human-AI co-creative systems. *ACM Transactions on Computer-Human Interaction, 30*(5), 1–28. https://doi.org/10.1145/3519026

Roschelle, J., & Teasley, S. (1995). The construction of shared knowledge in collaborative problem-solving. In C. O'Malley (Ed.), *Computer-supported collaborative learning* (pp. 69–97). Springer.

Sapienza, A., Goyal, P., & Ferrara, E. (2019). Deep neural networks for optimal team composition. *Frontiers in Big Data, 2*, 14. https://doi.org/10.3389/fdata.2019.00014

Sharples, M. (2023). *Towards social generative AI for education: Theory, practices and ethics*. https://doi.org/10.48550/ARXIV.2306.10063

Siemon, D. (2022). Elaborating team roles for artificial intelligence-based teammates in human-AI collaboration. *Group Decision and Negotiation, 31*(5), 871–912. https://doi.org/10.1007/s10726-022-09792-z

Wang, Q., Jing, S., Camacho, I., Joyner, D., & Goel, A. (2020). Jill Watson SA: Design and evaluation of a virtual agent to build communities among online learners. In *Extended abstracts of the 2020 CHI conference on human factors in computing systems* (pp. 1–8). https://doi.org/10.1145/3334480.3382878

Zheng, L., Long, M., Niu, J., & Zhong, L. (2023). An automated group learning engagement analysis and feedback approach to promoting collaborative knowledge building, group performance, and socially shared regulation in CSCL. *International Journal of Computer-Supported Collaborative Learning, 18*(1), 101–133. https://doi.org/10.1007/s11412-023-09386-0

Zheng, P., Li, S., Fan, J., Li, C., & Wang, L. (2023). A collaborative intelligence-based approach for handling human-robot collaboration uncertainties. *CIRP Annals, 72*(1), 1–4. https://doi.org/10.1016/j.cirp.2023.04.057

Chapter 11
AI Roles for the Team or Organization

11.1 Introduction

Consistent with the PKK model, in the HPL model, individuals in teams or organizations build a private world of knowledge on existing mental models and prior knowledge. Learners organize and relate new information to their existing understanding and for learning to be meaningful, content must be relevant, connected, and aligned with students' prior experiences. Importantly for social relevance and validation, learners must be engaged in activities that mirror the practices (e.g. the knowledge-in-action) of experts within a particular field. This involves immersing learners, for example, in *cognitive apprenticeships* (Resnick, 1984) of processes of inquiry, problem-solving, and collaborative discourse that experts use to build and refine their knowledge. Bransford underscores the idea that learning is not just about acquiring isolated facts but, as Dewey advocated, includes understanding how knowledge is interconnected and applicable in real-world contexts.

AI, when involved in a group or community's learning processes, can assist an expert community in its collective enterprise. For example, AI can work on behalf of the group to achieve collective goals, form sub-teams, analyze data, explore new areas for research, and can help suggest as well as manage changes that might take place in individual roles and responsibilities within the various teams and subgroups advancing the organization's mission. However, as noted in a recent report to the U.S. National Academy of Science (NAS et al., 2022), for the foreseeable future, AI will remain inadequate for autonomously operating in most novel situations because its performance may suffer due to unknown biases or limitations in training, or in the presence of challenges for which no clear "correct response" is known. This dependency on humans is vital in multi-team and multi-agent systems, such as those in a large organization like a university that may include AI in various places, at various times, and acting in a variety of roles.

© The Author(s), under exclusive license to Springer Nature Switzerland AG 2024

D. C. Gibson, D. Ifenthaler, *Computational Learning Theories*, Advances in Analytics for Learning and Teaching,
https://doi.org/10.1007/978-3-031-65898-3_11

This chapter will present a broad model based on the 'How People Learn' (HPL) framework introduced in Chaps. 9 and 10. The HPL frame of reference follows the implications of the socio-epistemological views and analysis of cognitive science and social constructivist theories of learning in social environments (Bransford et al., 2000); namely that learning takes place at the intersection of learners, community, assessment and knowledge.

11.2 The HPL Framework of a Learning Organization

Recall that the HPL model of the meso level has four nodes—the *learner* as a member of a social team or organization, the larger *learning community*, albeit a team, organization or loosely coupled community of practice, the *assessments* the community makes to admit, develop talent, and advance the interests of the community, the *expert knowledge* held by the members of the community. With these four network nodes in mind, this section presents some example concepts for utilizing AI to enhance a team or organization's interests. The HPL model was advanced in 2000 to explain the consensus opinion of cognitive scientists on how people learn; here we extend the model to how organizations learn, testing the idea that the processes and mechanisms that are good for individuals are also good for groups of individuals. There may be emergent ideas that arise when two or more people are learning and working together, so we next describe and compare two alternative models of a learning organization and then demonstrate the linkages to the HPL model.

Peter Senge of MIT and the Sloan School of Management presents five disciplines that are key to developing a learning organization (Senge et al., 1994): systems thinking, team learning, mental models, personal mastery and shared vision. *Systems thinking* involves understanding and representing how different components within a system interact and influence each other and encourages a holistic approach to problem-solving by developing a technical mastery of the interconnections and relationships between parts. *Team learning* encourages people to learn together, share insights, and collaborate to enhance the overall learning capacity of the entire team. *Mental models* are the deeply ingrained assumptions, beliefs, and generalizations that influence how individuals perceive and interpret the world. *Personal mastery* is about developing a high level of competence and a commitment to lifelong learning to achieve personal goals and contribute effectively to the organization. *Shared vision* fosters a sense of purpose and shared commitment, encouraging collaboration and coordination toward a collective vision for the future.

Extending on that foundation, David Garvin, of the Harvard Business School, emphasized that learning organizations possess certain capabilities, including the ability to create and transfer knowledge, the capacity to modify behavior based on new knowledge, and the skill to integrate new knowledge into everyday operations (Garvin, 1993). He highlighted the importance of continuous improvement as an ongoing commitment to enhancing processes, products, and services through the systematic gathering and application of new knowledge. Knowledge management, in his view, involves not only acquiring new knowledge but also effectively disseminating it throughout the organization, ensuring that insights are shared and applied across various departments and levels. Learning organizations also must encourage experimentation and take calculated risks, to learn from both successes and failures. Garvin also highlighted the importance of developing systematic processes for problem-solving and learning by analyzing past experiences, identifying patterns, and using these insights to inform future actions.

Table 11.1 compares the two learning organization frameworks (Senge and Garvin) with the HPL framework. As noted in a report for the US Army comparing four such frameworks (Snyder, 2016) commonalities among the models included aligning the organization's policies, systems, structure, and culture through the regular practice of learning and using learning to adapt. In the HPL model, such adaptation is the community evolving as its core knowledge and practices change, its membership changes, and new members join with new ideas that challenge the status quo while advancing the community's knowledge, practices and expertise.

The alignment in Table 11.1 is meant to be suggestive rather than definitive. The intent is to show that multiple theories can inform each other and add details that would otherwise remain tacit. For example, aligned with knowledge and assessment in the HPL model, is Senge's wholistic and rigorous systems thinking with a technical modeling language for documenting the archetypal dynamics of the community's epistemology (i.e. the scope of what the community knows, how it comes to know and validates what it knows) which also aligns with Garvin's systematic approach to solving problems. Another example is the alignment of Senge's mental models, defined as the generalizations and assumptions (including biases and past experiences mentioned by Garvin) from which we understand the world and act in it, and the interplay between individual and collective understandings in HPL.

In the sections that follow, the HPL framework is used to generate ideas for AI development in four domains: building and maintaining relationships with individuals, managing the adaptations of knowledge that keep the community's knowledge current, providing evidence-based feedback to members to assist in the

Table 11.1 Comparison of learning organization frameworks with HPL

Senge (1994)	Garvin (1993)	HPL (Bransford et al., 2000)
Systems thinking	Systematic problem solving	Knowledge & Assessment
Team learning	Transferring knowledge quickly and efficiently throughout the organization	Community
Mental models	Learning from past experiences	Learner & Community
Personal mastery	Learning from the best practices of others	Learner
Shared vision	Experimentation with new approaches	Knowledge & Assessment

development of expertise, and deepening current research while also finding and developing new lines of research.

11.2.1 Building and Maintaining Individual Learner Models

In the community version of the HPL model, individual learners are viewed as staff needing support and professional development to advance the community's expertise. The AI envisioned in this domain guides the development of inductees to provide leadership in research, knowledge sharing and community functions like assessment as well as maintaining the community.

11.2.1.1 Personalized Learning Paths

AI can construct and maintain individual learner models to aid in crafting personalized learning journeys that also promote the organizational mission and goals. From the perspective of a team or organization, the continuous learning of each member should be directed towards and contribute to the group's mission and effective group functioning through a combination of content knowledge, people-skills and access to appropriate technology. An effective pedagogical framework for this balance is based on Shulman's concept of teaching at the intersection of content and pedagogy (Shulman, 1987) that was expanded by Mishra & Koehler (2006) into a framework that included appropriate technology in both content and pedagogical arenas. An AI that aims to empower *adaptable directed individualized learning* toward a team's goals would answer these questions:

1. What technologies are authentically used by experts in the domain and what does this individual and the team need to know and be able to do with them to advance the team's mission?
2. What teaching situations and scenarios will call out, utilize and advance the knowledge and skills of this individual and team concerning the team's mission?
3. What domain area knowledge is needed and now seemingly missing from this individual and team's expressions, solutions and other responses?

11.2.1.2 Dynamic Model Updates

To ensure that the learner models evolve and adapt based on new skills, interests, or changes in performance balanced with the team's mission, AI could use the history of individual performance on this and other teams. If AI can help cluster people together to solve problems and address challenges, then it can also re-group, move people around mid-project and between stages, projects and even teams. For instance, AI could support finding someone with the right expertise to fill a temporary competence gap in a group activity or a compatible group of people for a short-term creative collaborative project within a larger project.

11.2.2 Managing Human Shaping of Knowledge Fields

Accelerating the productivity of science could be one of the most economically and socially valuable of all the uses of AI (OECD, 2023). In metadata discovery, a combination of humans and machines allows for better detection of scientific data artefacts. Soon, knowledge synthesis systems will continuously navigate the development of knowledge and integrate new submissions both from theory and empirical evidence. Intelligent interfaces could help experts find connections between concepts and theories cloaked in archaic academic language and emerging datasets and computational models (Chen & Hitt, 2021).

11.2.2.1 Content Curation

Artificial Intelligence generated content employs advanced generative AI techniques to decipher the intent of human input and generate appropriate content (Liu et al., 2023) that can assist in curating and organizing knowledge, fostering a coherent and structured knowledge landscape within the community. A framework for human-AI partnerships for content creation in an education study (Khosravi et al., 2023) utilized four key aspects: (1) *creating* novel content, (2) *evaluating* the quality of the created content, (3) *utilizing* learner-sourced contributions of students and (4) enabling *oversight* of the learner-sourcing process.

11.2.2.2 Knowledge Evolution Monitoring

AI can track knowledge gaps and emerging trends, assisting in reshaping the field of knowledge to meet community needs. Interdisciplinary interest in AI in knowledge management practices points to the expanding boundaries of research and its relevance to various academic disciplines (Akhavan et al., 2016). AI can detect and act on group process stages in knowledge creation. For example, AI can help identify when a group conversation is drifting away from its stated goals and objectives (Kovanović et al., 2016), helping assess the potential pros and cons of contemplated actions. AI can also help the team stay organized and productive by monitoring and characterizing individual contributions to collaboration (Kerrigan et al., 2019).

11.2.3 Providing Evidence-Based Feedback to Members

AI supported mentoring has been defined as a digitally enhanced and scalable process involving topics of a mentoring relationship including career support, support with expert knowledge, and socio-emotional support. In such processes, AI tools can take over and automate mentors' responsibilities, such as giving individual feedback or personal recommendations, helping to design individual learning paths or career strategies (Köbis & Mehner, 2021).

11.2.3.1 Performance Analytics

Performance-based analytics delivers personalized feedback based on performance data, offering insights for improvement and skill enhancement in AI-based coaching apps that are aiding supervisors in coaching people, for example AI can analyze recorded sales conversations and offer real-time coaching and feedback (Khandelwal & Upadhyay, 2021).

11.2.3.2 Behavioral Feedback

Before large language models, analysis of team behaviors in multi-agent settings was primarily conducted *post-hoc* and involved machine learning algorithms discovering multiple types of models of team behavior required for different granularities of team events from individual agent actions to social interactions to global team performance, for example (Nair et al., 2004) notes:

> It is sometimes beneficial to analyze the critical actions of single individuals because failures in these critical actions may have been extremely costly and hence correcting these failures can significantly improve performance. In other circumstances it is the collaborative agent interaction within a small sub-team that is key in team success or failure. Here, analysis may determine which sequences of agent interactions within the sub-team contribute to the team's success or failure. And in some other circumstances, an analysis of the global behavior trends of the entire team is important. (p. 70)

Team relationships in conjunction with task completion has recently been studied. In one series of studies, team working patterns were made visible to the instructor and teams, resulting in improved individual member behaviors (Kerrigan et al., 2019; Zamecnik et al., 2023). In another study the analysis approach used a bipartite network to support agent-to-agent and agent-to-task analysis in the same frame. In sum, AI can offer feedback on collaboration, communication, and engagement patterns, aiding in holistic team development.

11.2.4 Finding New Links and Research Lines

Rounding out the model structured by the HPL framework, AI can help the team or community find new relationships among its productions and documentation (which in some cases may already be functioning with international cross-disciplinary connections, as might be found in computational neuroscience, mathematics and medicine). In addition, with AI's increasing capabilities with inferences and big data, whole new lines of research may open via cross-domain connections and research recommendations.

11.2.4.1 Cross-Domain Connections

AI has already identified potential connections and interdisciplinary links within the field of medicine, fostering innovation and novel approaches. For example, a new structural class of antibiotics were recently discovered by applying ensembles of graph neural networks to extend from what was already known to predict antibiotic activity and cytotoxicity for 12,076,365 compounds (Wong et al., 2023) demonstrating that machine learning models can assist a team in research explorations. As another example of cross-domain connections, the refereed post-proceedings of an international workshop on Computational History and Data-Driven Humanities deals with the challenges and opportunities at the interface between computer science, social science, humanities, and mathematics (Bozic et al., 2016).

11.2.4.2 Research Recommendation Systems

One exciting area for AI application is in recommending new lines of research, new directions and areas of potentially fruitful exploration based on current trends and gaps in knowledge. To play this role, AI needs to engage in some part of the "complex problem-solving process, i.e., defining a problem, identifying root causes, proposing and evaluating solutions, choosing among options, making plans, taking actions, learning from past interactions, and participating in after-action reviews" as Seeber and colleagues note in their introduction (Seeber et al., 2020). AI can participate in distributed knowledge networks (e.g., building, searching, maintaining, and creatively accessing them to solve problems) by assisting teams in the problem-solving process, including exploring and understanding ideas, images, and semantics (Egozi et al., 2011; Ifenthaler, 2010), representing and formulating creative new representations, planning and executing team actions (Wang et al., 2021), and assisting a team in monitoring and reflecting on progress and achievement (Gao et al., 2018). Dynamic updating of multiple models operating at the macro level has been explored for tracing the development of research ideas globally and finding promising new avenues for research (Chubb et al., 2022). AI can thus be a creative and cooperating partner in collaborative research activity, making unique contributions in terms of ideas, processes, artifacts, and search results.

11.3 From Teams and Organizations to Global Enterprises

As a footnote for Chief Information Officers, additional information concerning the learning organization perspective on adopting AI comes from a report by the MIT Technology Review Insights, which surveyed 600 senior data and technology experts (Green, 2023). Findings included:

- Generative AI and LLMs are democratizing access to artificial intelligence, finally sparking the beginnings of truly enterprise-wide AI. AI is moving from pilot projects and "islands of excellence" to a generalized capability integrated into the fabric of organizational workflows.
- A trove of unstructured and buried data is now legible, unlocking business value will power extraordinary new advances across the organization.
- The generative AI era requires a data infrastructure that is flexible, scalable, and efficient and that democratize access to data and analytics, enhance security, and combine low-cost storage with high-performance querying.
- Some organizations seek to leverage open-source technology to build their LLMs, capitalizing on and protecting their data and IP.
- Automation anxiety should not be ignored, but on balance, the broader workforce will be liberated from time-consuming work to focus on higher value areas of insight, strategy, and business value.
- Unified and consistent governance are needed and critical given that generative AI brings commercial and societal risks, including protecting commercially sensitive IP, copyright infringement, unreliable or unexplainable results, and toxic content.

11.4 Limitations and Conclusions

This chapter intended to maintain the book's focus on learning processes while applying a framework from individual learning to teams and organizations, and further, aligning ideas for developing AI applications with organizational theorists who have promoted 'learning organizations.' Thus, several areas of organizational theory were ignored, for example, Bolman and Deal's approach (1991) that reframes organizations in terms of the leadership of structural, human relationships, political and symbolic arenas. AI will almost certainly play a role in every aspect of organizational leadership, but these possibilities are out of scope for the project to trace learning processes across scales and mechanisms from individuals to social groups, to society at large.

Eight recommendations concerning the development of AI that enhances team and organizational learning were presented, with links to exemplative literature. The eight concepts were organized by the four domains of HPL framework, a structure which was promoted as a fully connected network model of the meso level, namely: the learner as a team member, the learning community as a learning organization, the learning processes of admitting, inducting, and developing talent to advance the community's interests, and the growth and curation of specialized areas of knowledge. In each of these areas, AI not only can partner with humans but also join in the learning processes as an increasingly creative team member with some strengths that go beyond human capacities and some limitations that constrain its possibilities for contribution.

11.5 Moving Forward

At this point in the exposition of the theory, there is a fully described micro level network model of individual learning and a fully described meso level network of social learning. Each network has a limited set of nodes and relationships, but due to combinatorial considerations that arise from multiple copies as well as interpenetrating relationships, the number of analytical possibilities is quite large. At the individual level, since the network is a directed loop to represent causal relationships, there are fewer combinations to consider. However, since there might be many loops simultaneously present during a learning event and any combination of them might serve as the external inputs to another loop, the countability quickly becomes large and uncertain. At the social learning level, that large and uncertain status is multiplied by each member of a team or organization and in addition, the connections among the rest of the nodes at the meso level are multipoint, adding additional potential pathways of influence. Still, despite these difficulties in exactness, the network is bounded (it has exclusion principles) and is not infinite (it is countable), so with the assistance of AI, there is some hope of monitoring, shaping, modeling and thus better understanding how learning processes work in individuals and groups.

In the next section of the book, Chap. 12 and beyond, a well-known sociocultural framework known as Activity Theory (Engeström, 1999; Vygotsky, 1978) is re-introduced as a network for modeling learning processes at a cultural level, that is, where two or more learning organizations are interacting with each other and the larger world around them. In this macro level model, the meso and micro levels are subsumed, as the meso level of the learning organization is multiplied and emergent features are added. This mirrors the situation at the meso level, where the micro level individual unit was multiplied and when added with emergent features, resulted in a model of social learning.

References

Akhavan, P., Ebrahim, N. A., Fetrati, M. A., & Pezeshkan, A. (2016). Major trends in knowledge management research: A bibliometric study. *Scientometrics, 107*(3), 1249–1264. https://doi.org/10.1007/s11192-016-1938-x

Bolman, D., & Deal, T. (1991). *Reframing organizations: Artistry, choice, and leadership.* Jossey-Bass.

Bozic, B., Mendel-Gleason, G., Debruyne, C., & O'Sullivan, D. (Eds.). (2016). *Computational history and data-driven humanities* (Vol. 482). Springer International Publishing. https://doi.org/10.1007/978-3-319-46224-0

Bransford, J. D., Brown, A. L., & Cocking, R. R. (2000). *How people learn: Brain, mind, experience, and school.* http://www.nap.edu/openbook.php?isbn=0309070368

Chen, V. Z., & Hitt, M. A. (2021). Knowledge synthesis for scientific management: Practical integration for complexity versus scientific fragmentation for simplicity. *Journal of Management Inquiry, 30*(2), 177–192. https://doi.org/10.1177/1056492619862051

Chubb, J., Cowling, P., & Reed, D. (2022). Speeding up to keep up: Exploring the use of AI in the research process. *AI & SOCIETY, 37*(4), 1439–1457. https://doi.org/10.1007/s00146-021-01259-0

Engeström, Y. (1999). Activity theory and individual and social transformation. In Y. Engestrom, R. Miettinen, & R. Punamaki (Eds.), *Perspectives on activity theory* (pp. 19–38). Cambridge University Press.

Egozi, O., Markovitch, S., & Gabrilovich, E. (2011). Concept-based information retrieval using explicit semantic analysis. *ACM Transactions on Information, 0*(0), 1–38. https://doi.org/10.1145/2063576.2063865

Gao, J., Galley, M., Li, L., Brain, G., Brockett, C., Celikyilmaz, A., Cheng, Y., Dolan, B., Fung, P., Gan, Z., Lee, S., Li, J., Li, X., Liu, B., Madotto, A., Majumder, R., Papangelis, A., Pietquin, O., Quirk, C., ... Zhang, Y. (2018). Neural approaches to conversational AI question answering, task-oriented dialogues and social chatbots. arXiv.Org.

Garvin, D. (1993). Building a learning organization. *Harvard Business Review, 71*(4), 78–91.

Green, A. (2023). *The great acceleration: CIO perspectives on generative AI* (MIT Technology Review Insights). MIT.

Ifenthaler, D. (2010). Learning and instruction in the digital age. *Learning, 3–10.* https://doi.org/10.1007/978-1-4419-1551-1

Kerrigan, S., Feng, S., Vuthaluru, R., Ifenthaler, D., & Gibson, D. (2019). Network analytics of collaborative problem-solving. In *Proceedings of the 16th international conference on cognition and exploratory learning in digital age (CELDA 2019)* (pp. 43–50). https://doi.org/10.33965/celda2019_201911L006

Khandelwal, K., & Upadhyay, A. K. (2021). The advent of artificial intelligence-based coaching. *Strategic HR Review, 20*(4), 137–140. https://doi.org/10.1108/SHR-03-2021-0013

Khosravi, H., Denny, P., Moore, S., & Stamper, J. (2023). Learnersourcing in the age of AI: Student, educator and machine partnerships for content creation. *Computers and Education: Artificial Intelligence, 5*, 100151. https://doi.org/10.1016/j.caeai.2023.100151

Köbis, L., & Mehner, C. (2021). Ethical questions raised by AI-supported mentoring in higher education. *Frontiers in Artificial Intelligence, 4*, 624050. https://doi.org/10.3389/frai.2021.624050

Kovanović, V., Joksimović, S., Waters, Z., Gašević, D., Kitto, K., Hatala, M., & Siemens, G. (2016). Towards automated content analysis of discussion transcripts. In *LAK '16: Proceedings of the sixth international conference on Learning Analytics & Knowledge* (pp. 15–24). https://doi.org/10.1145/2883851.2883950

Liu, G., Du, H., Niyato, D., Kang, J., Xiong, Z., Kim, D. I., Xuemin, & Shen. (2023). *Semantic communications for artificial intelligence generated content (AIGC) toward effective content creation.* https://doi.org/10.48550/ARXIV.2308.04942

Mishra, P., & Koehler, M. J. (2006). Technological pedagogical content knowledge: A framework for teacher knowledge. *Teachers College Record, 108*(6), 1017–1054. https://doi.org/10.1111/j.1467-9620.2006.00684.x

Nair, R., Tambe, M., Marsella, S., & Raines, T. (2004). Automated assistants for analyzing team behaviors. *Autonomous Agents and Multi-Agent Systems, 8*, 69–111.

NAS, Committee on Human-System Integration Research Topics for the 711th Human Performance Wing of the Air Force Research Laboratory, Board on Human-Systems Integration, Division of Behavioral and Social Sciences and Education, & National Academies of Sciences, Engineering, and Medicine. (2022). *Human-AI teaming: State-of-the-art and research needs* (p. 26355). National Academies Press. https://doi.org/10.17226/26355

OECD. (2023). *Artificial Intelligence in science: Challenges, opportunities and the future of research*. OECD. https://doi.org/10.1787/a8d820bd-en

Resnick, L. B. (1984). Education and learning to think. In *Learning*. Falmer Press.

Seeber, I., Bittner, E., Briggs, R. O., De Vreede, T., De Vreede, G.-J., Elkins, A., Maier, R., Merz, A. B., Oeste-Reiß, S., Randrup, N., Schwabe, G., & Söllner, M. (2020). Machines as teammates: A research agenda on AI in team collaboration. *Information & Management, 57*(2), 103174. https://doi.org/10.1016/j.im.2019.103174

Senge, P. (1994). *The fifth discipline: The art and practice of the learning organization*. Doubleday.

Senge, P., Kleiner, A., Roberts, C., Ross, R., & Smith, B. (1994). *The fifth discipline fieldbook: Strategies and tools for building a learning organization*. Currency Doubleday.

Shulman, L. S. (1987). Knowledge and teaching: Foundations of the new reform. *Harvard Educational Review, 57*, 1–22.

Snyder, J. (2016). *Learning organization models and their application to the U.S. Army* (p. 30). U.S. Army Research Institute for the Behavioral and Social Sciences. https://apps.dtic.mil/sti/tr/pdf/AD1036965.pdf

Vygotsky, L. S. (1978). *Mind in society: The development of higher psychological processes*. In M. Cole, S. John-Steiner, S. Scritmer, & E. Souberman (Eds.), Harvard University Press.

Wang, Y., Zhong, F., Xu, J., & Wang, Y. (2021). *ToM2C: Target-oriented multi-agent communication and cooperation with theory of mind*. https://doi.org/10.48550/ARXIV.2111.09189

Wong, F., Zheng, E. J., Valeri, J. A., Donghia, N. M., Anahtar, M. N., Omori, S., Li, A., Cubillos-Ruiz, A., Krishnan, A., Jin, W., Manson, A. L., Friedrichs, J., Helbig, R., Hajian, B., Fiejtek, D. K., Wagner, F. F., Soutter, H. H., Earl, A. M., Stokes, J. M., et al. (2023). Discovery of a structural class of antibiotics with explainable deep learning. *Nature*. https://doi.org/10.1038/s41586-023-06887-8

Zamecnik, A., Kovanovíc, V., Joksimovíc, S., Grossmann, G., Ladjal, D., Marshall, R., & Pardo, A. (2023). Using online learner trace data to understand the cohesion of teams in higher education. *Journal of Computer Assisted Learning*, jcal.12829. https://doi.org/10.1111/jcal.12829

Chapter 12
A Network Theory of Culture

12.1 Introduction

In Vygotsky's theory, social interaction plays a fundamental role in the development of cognition on two levels: first, through *interpsychological* interaction with others at the social level, and second, integrated into the individual's *intrapsychological* mental structure (Vygotsky, 1978). This accords well with the picture drawn thus far of the network model representations of the micro and meso levels. In the preceding chapters, a *causal learning loop* model of the intrapsychological mechanisms of mental structure was discussed as a network representation of micro level learning processes. A *fully connected (complete digraph) network* model of interpsychological social learning interactions was discussed as a network representation of meso level learning processes.

The macro level network is introduced in this final section of the book, attributable in part to Vygotsky, who introduced the concept of *cultural mediation*, emphasizing the role of cultural tools and artifacts in mediating human cognition and behavior. Cultural mediation of intelligence and behavior occurs because humans act collectively, learn by doing, and communicate in and via their actions for example, by making, using, and improving tools including symbols and now, smart tools, to learn and communicate, and people create these symbols (e.g. language, gestures, art, smart tools) and interpret their meaning within communities (Vygotsky, 1978).

12.2 Cultural Mediation

Based on the idea of cultural mediation, Yrjö Engeström, a scholar within the cultural-historical tradition developed and expanded on Vygotsky's idea of artifacts mediating the relationship between subjects (e.g. individuals at the micro level or

© The Author(s), under exclusive license to Springer Nature Switzerland AG 2024

D. C. Gibson, D. Ifenthaler, *Computational Learning Theories*, Advances in Analytics for Learning and Teaching,
https://doi.org/10.1007/978-3-031-65898-3_12

groups at the meso level) and objects (e.g. learning aims, goals or outcomes) by introducing the concept of an *activity system* (Engeström et al., 1998).

Activity theory thus views human activities as complex systems involving multiple individuals, artifacts, and rules. An activity system can be viewed as a fully connected network (i.e. a complete undirected graph where all the edges are bidirectional) of six elements: subjects, that is the actors and agents involved, objects or goals, aims, objectives as well as outcomes of the activity, tools and artifacts used in the activity, rules that provide explicit and implicit guidelines, community as the social context and with all the functions outlined at the meso level, and division of labor e.g. how tasks are distributed (Fig. 12.1). The macro level network thus has six nodes defined below and as a complete digraph with self-loops, allows many possible trajectory sequences from a base of 36 relationships (e.g. each node can connect to 5 other nodes plus itself, so the graph has 6*6 edges).

The complete graph representation of culture as an activity system assists with defining and understanding the specific interactions, tensions and influences among the elements that are crucial for understanding how a particular event or process arises and evolves. All links are two-way, which allows for both self-reflection and self-referential loops in the possible trajectories. The nodes have been extensively discussed in the scholarship of cultural historical activity theory (Engeström et al., 1998) and as a framework for pedagogy (Jonassen & Rohrer-Murphy, 1999), but less so as a computational modeling framework.

Artifact node includes tools, systems, symbols, language, devices, as well as other computationally enhanced cognitive extensions and intelligence distributions.

Object is the purpose of using a mediating artifact to accomplish an aim or goal.

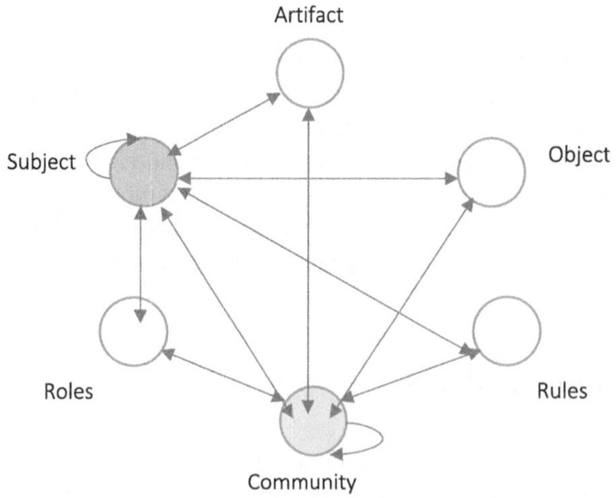

Fig. 12.1 Cultural historical activity theory elements. Note: A dynamic CHAT analysis from the perspective of a learner would consider 6 influence edges changing over time, while an analysis from the perspective of a cultural community would consider a different set of 6 relationships changing over time

Subject is the person, agent or agency and is synonymous with the HPL meso Learner and subsumes one or more Piaget-Kauffman-Keller causal learning cycles.

Roles differentiate community members who inhabit specialized niches of performance according to their agency or expertise.

Community includes and expands on the single HPL community of experts into a larger conglomerate global, historical, multicultural, geosocial world.

Rules include membership rules, norms, and practices of the activity system community, rules of entry into the community and the rule of entry of new knowledge into accepted fields.

An event (e.g. two expert communities give rise to a new interdisciplinary field of study) or process (e.g. how research is conducted and compared in each field and the new interdisciplinary field) can be understood as a trajectory, a path over time, visiting the nodes and traversing the relationships of the macro network. When needed, an analysis may also cross into the nodes and relationships of the meso or micro levels. Such a trajectory, which may involve both directed and cyclic subnetworks, is addressable by a variety of AI methods focused on causal modeling (Xiong, 2022). For example, a combination of reinforcement learning, bidirectional conditional generative adversarial networks, transformers and attention techniques "can be used to construct large functional causal networks" (p.xxi).

Engeström also introduced the notion of *expansive learning* which can be compared with Kauffman's idea of expanding into the *adjacent possible* at the micro level (Chap. 6). In activity theory, expansion is required when there is a contradiction, disparity or incongruence in any of the six elements, leading to a need for change and adaptation. Note the similarity of the idea of contradiction to *disequilibration* giving rise to attention in the causal learning cycle (Chap. 4). Also note that expansion through adaptation is consistent with Piaget's idea of *accommodation* and equates with Holland's *quasi-homomorphism* mechanism of forming exception rules in a massively parallel and dynamical control system (Holland, 1998), as discussed at the micro level.

In Holland and Kauffman's terms, expansion into the adjacent possible is accomplished via additional fine graining and elaboration within a boundary (e.g. the boundary of the brain in an individual, the boundary of a learning community's limits of knowledge in a team or organization, and the boundary of the global community in a complex activity system). Recall the toy model of a bounded infinity introduced in Chap. 1, which illustrates how an infinite continuum can exist within a well-defined and bounded interior of an open system via self-referential loops. The fine-grained internal elaboration of private, public and global experience creates specific contexts, exception rules and new sensitivities at all three levels (e.g. individual, social, cultural).

The process of expansion and accommodation at the macro or cultural level, according to Engeström, involves negotiation, reflection, and the creation of new tools or rules to address the challenges faced within the activity system, a learning process that in the case of an individual, Piaget would identify as the mechanism of accommodation and Kuhn, speaking of the evolution of ideas in a community, would say is the inflection point of a paradigm shift in knowledge. In summary, Engeström's updating of Vygotsky's theory which became cultural-historical activity theory emphasizes both the collective context of the emergence of intelligence and the dynamic nature of human activities like learning.

12.3 Relating the Macro to Meso Levels

Often, levels above another take inputs from lower levels, summarize and cluster them into new sparse representations, and use them to predict and influence actions. However, the precedence can be reversed when an existing schema, idea or other representation is the primary or sole source for creative elaboration, as when an individual dreams or imagines, a team or organization creates a vision or mission, and a culture evolves with new symbols, rituals and meaning (Stein, 1994).

In the book *The Dream of Culture*, Stein…

> …explores the hypothesis that culture is our dreaming together while we are wide awake. People in groups, awake, do with their shared symbolic and ritual materials, their technology and their environment, what we each also do while asleep: wish what we may dare, and disguise our wishes; contrive our forbidden victories and stage our defeats; attempt to master what we have found overwhelming. Culture and dreams are part of each other, twin (isomorphic) facets of the same process. In this book, Howard Stein explores the work of culture as if it were a species of dream work (Stein, n.d.).

Entities in *dynamic CHAT network theory* relate to the HPL and PKK models in clusters as well as through *subsumption architecture* in which more complex upper layers not only depend on lower, more reactive layers, but can also influence their behavior (Brooks, 1991).

Artifact joins with **Rules** and **Objects** to form the *HPL Knowledge*.
Subject (*HPL Learner*) joins with **Community** and **Roles** to form *HPL Assessment*.

In Fig. 12.2 the 3-level model combining the micro, meso and macro levels as first introduced in (Gibson et al., 2023) is illustrated.

The *causal learning cycle* at the micro level activates individual learning and may do so in parallel as well as in hierarchical structures within the individual. Taken as a sum or current state, the learner interacts with the social world at the meso level and may be a member of a team or organization, where community norms, processes and knowledge bind the members together in facing challenges, solving problems or exploring the world. The interactions and current states of the nodes at the meso level are driven by the same motive forces found at the micro level. That is, discontinuities of any kind, at any node of the HPL model (i.e. *knowledge, community, assessment or learner*) causes *attention, assimilation, relevance, accommodation, confidence, equilibration and satisfaction* processes in the local social milieu. As those processes or events transpire, they trace trajectories in the meso space, just like individual learning processes trace trajectories in the micro

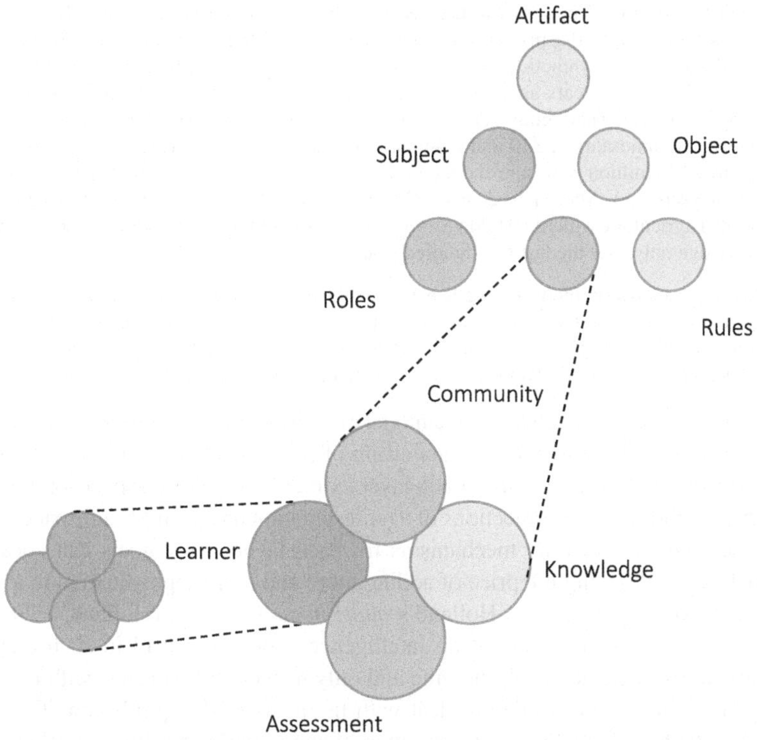

Fig. 12.2 3-level model of learning processes. Note: Micro level causal learning cycles provide information and influence viz. the Learner, Meso level HPL framework provides information and influence in an expert community, Macro level elements provide context information and influence for cultural evolution, community roles in culture and individual learning.

space. If the events and processes of the meso level impact any other community (or any other macro system element) then the macro environment becomes engaged, and once again unique trajectories wend through the macro space, bounded by both time and the remaining nodes and relationships of the 3-level network model. The subsumption architecture supports learning simultaneously at all three tiers as well as the parallel processing of multiple trajectories within and across the tiers, while retaining knowledge of the past and building new *structural causal models* (Xiong, 2022) in the appropriate agent. Recall that the agent at the micro level is the individual, at the meso level is the team or organization and at the macro level is the encompassing culture.

Rodney Brooks introduced the idea of the subsumption architecture to counter a prevailing notion that a central processing unit or central representation of the world was needed for intelligence. The essence of his idea was that intelligence could be built up from layers of task-achieving behaviors supported by real-world feedback. He arrived at this position by thinking about how biological evolution had spent its time leading up to human intelligence. From his paper *Intelligence without representation* (Brooks, 1991).

> It is instructive to reflect on how earth-based biological evolution spent its time. Single-cell entities arose out of the primordial soup roughly 3.5 billion years ago. A billion years passed before photosynthetic plants appeared. After almost another billion and a half years, around 550 million years ago, the first fish and vertebrates arrived, and then insects 450 million years ago. Then things started moving fast. Reptiles arrived 370 million years ago, followed by dinosaurs at 330 and mammals at 250 million years ago. The first primates appeared 120 million years ago and the immediate predecessors to the great apes a mere 18 million years ago. Man arrived in roughly his present form 2.5 million years ago. He invented agriculture a mere 10,000 years ago, writing less than 5000 years ago and "expert" knowledge only over the last few hundred years.
>
> This suggests that problem solving behavior, language, expert knowledge and application, and reason, are all pretty simple once the essence of being and reacting are available. That essence is the ability to move around in a dynamic environment, sensing the surroundings to a degree sufficient to achieve the necessary maintenance of life and reproduction.

The subsumption approach leads to a model of distributed intelligence in which layers of an active sensing-acting system are aware of (e.g. pay attention to, react to, monitor) a subspace of the real world. Each layer extracts "only those aspects of the world which they find relevant-projections of a representation into a simple subspace" (p. 6). Because there are separate mechanisms for each layer, many goals can coexist in parallel, without paying the price of adding more and more sophistication to a single representation and processor. Holland's *quasi-homomorphism* (Holland, 1998) provides a mechanism for building up intelligence about the world in such a system because all rules are active all the time and only new exception rules, with increased energy investment, are needed to deal with moments of disequilibrium. The world model is stable and quiet, using a minimum of energy until learning is needed.

At the micro level, Hawkin's proposal that the neocortex has multiple copies of such mechanisms allows for the continuous elaboration of experience into finer and finer grained knowledge of the world (Hawkins, 2021). The causal learning cycle at the micro level describes the four stages of the elaboration process, which propels the individual into Kauffman's *adjacent possible*, also known concerning individual learning as Vgotsky's *zone of proximal development.*

At the meso level, fine grained elaboration is bounded by the functional elements of Learner, Community, Knowledge and Assessment. Learners can be added to and subtracted from the expert community over time, and any one individual can bring a new idea that adds to or upsets a generation of prior knowledge. Knowledge can grow more sophisticated, shrink in importance, or get re-organized and re-framed with new paradigms. Assessment, which is how the community senses and comes to know and act, knows when it knows what to do and what to think, and what to say to neophytes, acolytes and experts, can evolve new measures, methods and analytics. Community itself can evolve, disappear, merge with other communities, or stand in contrast, as needed over time.

At the macro level, each node of the network is a bounded infinity influencing the dynamics of the whole and all its subspaces as well as its interpenetrations with the meso and micro levels. Any point of analysis can treat all the remaining nodes of the network as context either within a level, between two levels or considering the entire 3-tiered framework (Table 12.1).

Table 12.1 Example units of analysis, aims & context and example research questions

e.g. Units of analysis	e.g. Aims & contexts	e.g. Research questions
Micro Level Targets		
PKK Disequilibration	Impact on balance of Piaget-Keller-Kauffman nodes and relationships.	To what extent does an intervention cause an effect in one or more balances of the PKK model (e.g. the learner's attention, assimilation, sense of relevance, etc.)? Do the 'effect probabilities' tail off as predicted by the causal loop model (e.g. are probabilities stronger the closer the impacted element is to the unit of analysis)?
PKK Disequilibration	Impact on external context of an HPL (Change in HPL nodes or relationships) where the Learner is embedded.	To what extent does the PKK variable impact the HPL node(s). Is there a relationship between the node or relationship of the PKK to one or more nodes and relationships in the HPL?
PKK cycle or arc speed, intensity or scale	Learner's impact on an HPL expert community (the knowledge, assessment, community).	To what extent do full cycles of the PKK (or subset arcs) impact on the Learner's role in an HPL event or process?
Meso Level Targets		
Learner	Trajectory toward mastery.	What is the distribution of nodes and relationships and the trajectory characteristics (e.g. pathlength, nodes included, time of traversal) given PKK cycle or arc characteristics of a Learner?
Learner	Influence of Learner's PKK on the HPL elements of an organization.	Focusing on changes to Knowledge, Assessment and Community, to what extent was a particular Learner influential and to what effect?
Community	Paradigm shifts in knowledge given changes in the remaining HPL elements.	To what extent did the holdout HPL element(s) influence the change? Using the four elements, what node was involved in the initiation of the change and what were the trajectory characteristics (significant nodes and relationships, time of traversal) of the change?
Assessment	Paradigm shifts in methods, measures and validation processes given changes in the remaining HPL elements.	To what extent did the holdout HPL element(s) influence the change?
Knowledge	Paradigm shifts in knowledge given changes in the remaining HPL elements.	To what extent did the holdout HPL element(s) influence the change?

(continued)

Table 12.1 (continued)

e.g. Units of analysis	e.g. Aims & contexts	e.g. Research questions
Macro Level		
Artifact	Technology capability that drives cultural or community shift.	What is the HPL description of the artifact (e.g. what knowledge community created it, how is the technology used in knowledge production, assessment and community building).
Object	Impact of Learner's aims on self, team and classroom community. Impact of HPL element's aim on a new Learner. Impact of culture on an HPL community of practice.	Influences on a team's aims from an individual.
Rules	How the PKK characteristics of one person influences a team or organization's rules. How rules changes impact an HPL organization or a cooperation among HPL organizations.	How diffusion of influence differs in an HPL team or organization based on the PKK characteristics of its members.
Subject	How PKK and Learner characteristics impact on Object and Artifact.	Apply Micro and Meso level findings to an individual within a cultural setting (e.g. classroom, school, community).
Role	What influences role performance and role changes from the remaining factors of Activity Theory.	Conduct knock-out research on the nodes and links of the Macro Level.
Community	What are the HPL characteristics of an organization that lead it to collaborate with other HPL organizations?	How drivers differ as the initiation of change varies from technology, knowledge, assessment, or community norms.

12.4 Limitations and Conclusions

Since the theory in this chapter is proposed as a macro-level representation, it may not fully account for some unique and specific cultural practices and learning processes that exist via macro level influences on the meso and micro levels. The scientific and humanistic process of identifying contradictions, disparities, or incongruences and enacting change and adaptation within cultural learning environments may be more complex and context-dependent than the macro level model can predict. The fullness of criticism and rejoinders concerning cultural historical activity theory have not been sufficiently represented in the chapter to refute the idea that some influence on culture may not be captured by the 6-node macro level model. In (Roth & Lee, 2007) however, the CHAT Theory was asserted to be a 'neglected legacy' of Vygotsky which the intervening years have in large measure rectified, but a full literature review is beyond the scope of the focus on modeling and AI interventions into learning processes.

Six primary concerns limiting the theory from its foundations were raised and discussed by the founders (Vygotsky, Leontiev and Engeström) but others have commented and should be mentioned. First, the issue of complexity and lack of clarity has been discussed by Michael Cole as a need to include culture as a context and empirical methods that begin with a concrete activity system. In the current project of the 3-tiered model, a 'concrete activity system' is one in which there are streaming data collecting sensors for as many of the nodes and relationships as possible, collecting time-stamped state information concerning the states of the nodes and flows of information across the network. As this capability was not available at the foundation of CHAT Theory, nor was there a model of the meso and micro level interpenetrating with the theory, the theory was historically less computationally relevant than it can be today.

Second, the deterministic implications of the emphasis on social and historical factors influencing human activity may lead to an oversimplified view that downplays individual agency and the potential for personal initiative. Despite the role of the 'Subject' in the theory, both Leontiev and Vygotsky shared the concern. The 3-level model addresses this concern with two additional levels of analysis, bringing new focus to the agency of an individual and the potential impacts of individual agency on a team or organizational context and reciprocal impacts back to the individual and the micro-to-meso also linking to the macro level. The 3-level model's emphasis on dynamic equations driving network states and a hierarchical framework allows individual agency to influence upward from the micro into the meso and macro levels, carrying with it embedded causality; and reciprocally allows a cultural influence to influence an organization or team, including its members.

Third, inadequate attention to individual development was a shared concern of the founders. Here too, the issue is addressed by the 3-level model's emphasis on interpenetrating influences that cross the boundaries of micro-to-meso-to-macro leaving a computable trajectory in the space of possibilities.

Fourth, the original CHAT theory can be criticized for its rather static conception (Cole, 1999), a problem that surfaces in all research that ignores time and evolutionary processes. The 3-level model emphasizes that all analyses *must* include time and history and remain focused on modeling and intervening in learning *processes.*

Fifth, the original CHAT theory has been criticized for a lack of practical guidance, but perhaps its contribution has been rather to organize theory and provide a framework for analysis. For example, Annalisa Sannino has promoted Vygotsky's call for an indirect method in psychological investigations to advance psychological theory and methods (Sannino & Sutter, 2011). While examples abound in using the framework for sociocultural analyses, more to the point of this book, it has also appeared in research on the impacts of technology in education such as in Sasha Barab and colleagues publications (Barab et al., 2002, 2004).

Sixth, with CHAT's origins in Soviet psychology and its subsequent development in western academia have led to concerns about its cultural bias. Some critics argue that the theory may not be universally applicable and might not fully account for cultural variations in all human activity. This criticism, to be relevant for this book's thesis, should call forth counter examples in which a cultural shift or influence on a learning process has made an impact on an individual, a team, organization or a larger cultural unit and is not able to be explained, simulated or explored by the 3-level network model. Without such evidence to weigh against, the current model may limited, but is more likely simply under-supported by current empirical evidence due to omissions of its features to date (e.g. the requirement of the inclusion of time, evolutionary processes, and a computational representation space analyzed with assistance of machine learning tools, fed by a multilevel and multimodal sensor net).

The selection of units of analysis within and across levels, while providing a flexible framework, also suggests a limitation in terms of the potential for overlooking important cultural elements that do not fit neatly into the predefined network nodes and relationships. For example, an individual's family and sociocultural background may have features that at first look do not appear to be related to the meso level (for family) and macro level (for cultural markers). This implies that every 3-level analysis should determine specific weights for every node of the complete model, which has not been explored in detail in this narrative and is left to the reader to contemplate. It is nevertheless possible and feasible that the structural causal model as presented in high detail using many of the leading edge ideas of AI as of 2022 in (Xiong, 2022) has sufficient tools for representing and simulating any trajectory of the 3-tiered system of learning processes. [possibly develop an appendix with more details of the SCM.]

With these limitations in mind, this chapter aimed to provide a theoretical foundation for practical applications that can enhance learning processes at the cultural level via AI applications developed with an understanding of cultural historical activity theory. As the macro level of the tri-level model of learning processes, the *dynamic CHAT* model's elements (as nodes) and relationships (as edges) provide a network framework for understanding the macro level's interpenetration with the micro-meso levels and implies a network theory of culture that suggests a robust computational approach is possible for analyzing cultural phenomena in learning environments.

In discussing the interpenetration of the macro with the meso and micro levels, this chapter also touches upon the essence of a multilevel AI agency (including groups of agents) coevolving in a dynamic environment with a shared purpose of enhancing learning processes at every level. The coevolutionary perspective can help ensure the development of AI tools and interventions that are sensitive to the complexities of cultural dynamics in learning environments. It implies that AI applications can model complex cultural dynamics and contribute to the development of interventions that can positively influence learning outcomes at the cultural level.

12.5 Moving Forward

In the next chapter, the ideas presented at the macro level are used to envision AI applications that intervene and advance learning processes at the cultural level.

References

Barab, S., Barnett, M., Yamagata-Lynch, L., Squire, K., & Keating, T. (2002). Using activity theory to understand the systemic tensions characterizing a technology-rich introductory astronomy course. *Mind, Culture, and Activity, 9*(2), 76–107. https://doi.org/10.1207/S15327884MCA0902_02

Barab, S., Schatz, S., & Scheckler, R. (2004). Using activity theory to conceptualize online community and using online community to conceptualize activity theory. *Mind, Culture, and Activity, 11*(1), 25–47. https://doi.org/10.1207/s15327884mca1101_3

Brooks, R. A. (1991). Intelligence without representation. *Artificial Intelligence, 47*, 139–159. https://doi.org/10.1016/0004-3702(91)90053-M

Cole, M. (1999). Cultural psychology: Some general principles and a concrete example. In Y. Engeström, R. Miettinen, & R.-L. Punamäki (Eds.), *Perspectives on activity theory* (1st ed., pp. 87–106). Cambridge University Press. https://doi.org/10.1017/CBO9780511812774.008

Engeström, Y., Miettinen, R., & Punamäki, R. (1998). *Perspectives on activity theory*. Cambridge University Press.

Gibson, D., Kovanovic, V., Ifenthaler, D., Dexter, S., & Feng, S. (2023). Learning theories for AI promoting learning processes. *BJET*. https://doi.org/10.1111/bjet.13341

Hawkins, J. (2021). *A thousand brains: A new theory of intelligence*. Basic Books.

Holland, J. (1998). *Emergence: From chaos to order*. Helix Books Addison Wesley/Perseus Books Group.

Jonassen, D., & Rohrer-Murphy. (1999). Activity theory as a framework for designing constructivist learning environments. *ETR&D-Educational Technology Research and Development, 47*(I), 66–79.

Roth, W.-M., & Lee, Y.-J. (2007). "Vygotsky's neglected legacy": Cultural-historical activity theory. *Review of Educational Research, 77*(2), 186–232. https://doi.org/10.3102/0034654306298273

Sannino, A., & Sutter, B. (2011). Cultural-historical activity theory and interventionist methodology: Classical legacy and contemporary developments. *Theory & Psychology, 21*(5), 557–570. https://doi.org/10.1177/0959354311414969

Stein, H. F. (1994). *The dream of culture: Essays on culture's elusiveness*. Psyche Press.

Stein, H. (n.d.). *The dream of culture* [Online Library]. Library of Social Science. https://www.libraryofsocialscience.com/books/the-dream-of-culture/

Vygotsky, L. S. (1978). *Mind in society: The development of higher psychological processes*. In M. Cole, S. John-Steiner, S. Scritmer, & E. Souberman (Eds.), Harvard University Press.

Xiong, M. (2022). *Artificial intelligence and causal inference* (1st ed.). CRC Press, Taylor & Francis Group.

Chapter 13
AI Roles in Cultural Learning

13.1 Introduction

The term 'cultural learning' distinguishes the macro level's learning processes from those of a singular team or organization at the meso level. The assumption at the meso level is that the mission and aims of a team or organization sets the goal for member activities, including AI as a participating member. Any conflicts between team and individual goals must be resolved to the benefit of the team or organization's aims, otherwise the social fabric frays. At the macro level, more than one team or organization may exert influences on each other, so goal conflict resolution is more complex. Where a meso level mission is an internal organizing directive, at the macro level the mission (e.g. Object in Vygotsky's terminology) is an external influence on another team or organization as well as on the individuals who are members. In the 3-level model of learning processes, dynamics, irreversible time and the language and methods of graph and network theory are added to Vygotsky's and Engëstrom's theories to create a computationally relevant modeling framework with defined pathways of influence coming into and going out from the meso and micro levels. The network representation of CHAT will be titled the *dynamic CHAT* model.

13.2 AI as a Mediator and Initiator of Cultural Shifts

To help initiate and support cultural learning and shifts, AI at the macro level needs to encompass, leverage and delimit the embedded influences from micro and meso levels, while focusing on knitting together the separate and potentially conflicting communities of inquiry that might not otherwise find each other or work together. Practical examples of research methods enhanced by AI at this level include

© The Author(s), under exclusive license to Springer Nature Switzerland AG 2024

D. C. Gibson, D. Ifenthaler, *Computational Learning Theories*, Advances in Analytics for Learning and Teaching,
https://doi.org/10.1007/978-3-031-65898-3_13

computational approaches to bibliographic network analysis that may lead to new lines of research (Lim & Buntine, 2016), tracing the global spread of ideas (Molontay & Nagy, 2021), and finding solutions for out-of-field problems (Castelvecchi, 2024). Time is a critical dimension in many cases as there is often a significant amount of time between discovering new ideas and their practical application, typically only after society has evolved to be able to use those ideas, as is sometimes the case with mathematics (Feng et al., 2017). The mechanics within the macro level as well as its interpenetrating dynamics with the meso and micro levels can be modelled by deep learning methods and a variety of other machine learning models (Xiong, 2022).

Boundary crossings between the micro-meso-macro levels of the system carry replicated submodels or *persistence homologies* (Aktas et al., 2019) of the lower levels. For example, entering the meso level are multiple copies of learners with their micro processes of learning, and entering the macro level are multiple copies of teams and organizations with various missions and aims. Some dynamic influences are expected to persist from the micro to the meso level, such as individual learner characteristics known to influence group roles. At the meso level, each community of practice or field of knowledge and practice influences upward into the macro level and downward into the micro level. Whether the influences from one level to another are dynamic summaries (e.g., running averages within time windows, phase portraits, vector space as a context) or generative sparse representations (e.g., used for prediction, elaboration, sensor net construction) depends upon many factors including the node's current role in a function, the context of the function, and the causal role of the function.

The 3-level theory expands on the typical idea of 'context' by fully defining the boundary of context as the remaining nodes in the remaining levels from any entity and its level in the network. To support this conception of context, the theory requires that each node of the relevant network at each level has semipermeable boundaries resulting in them being polysemantic nodes to allow multiple-model participation (i.e., probabilistic composition context information comes from elsewhere in the hierarchy and influences are both received by and sent to other nodes or levels). The neuroscience concept for this phenomenon is 'reentrant processing' (Edelman & Mountcastle, 1982) and has been discussed in conjunction with the ontogeny and embedded nature of intelligence in works such as (Baum, 2003; Braitenberg, 1984; Pfeifer et al., 2007).

13.3 Goals for AI Assisting Cultural Learning

There six broad goals that stem from considering how AI can enhance cultural learning by focusing on the role of the nodes defined by the *dynamic CHAT* framework (Table 13.1). AI can not only provide new tools to individuals and organizations for the creation of artifacts, but can make original contributions and novel suggestions concerning the creation, application and semantic import (Langer,

Table 13.1 The macro level of sociocultural evolution: Cultural historical activity theory (CHAT) enhanced by AI

Entities in CHAT network theory	The focus of AI systems co-participating in cultural learning and intelligence via *dynamic* CHAT
Artifact—includes tools, systems, symbols, language, devices, and in general, computationally enhanced cognitive extensions and intelligence distribution.	AI **intelligent tools and resources generate novel ideas** for creating narratives, rationales, empirical tests, visualizations, and other learning artifacts.
Object—an aim or goal of using a mediating artifact to accomplish a goal or achieve an outcome.	AI **enhances goals and planning** to clarify objectives and find routes toward achieving stated aims.
Subject—the individual 'self' or agent directing learning that is synonymous with the Bransford meso *Learner* and the Piaget-Kauffman-Keller micro *learning agent*.	See AI roles in the meso and micro levels above.
Community—includes and expands on HPL *Community* into a larger global and historical, multicultural, geosocial world. The global community of a field of knowledge is an agency unto itself when it is a field (community of practice) relating to the rest of the world.	AI **helps cross-cutting communities form and evolve** and helps the expert community **integrate with other fields of knowledge** via innovation and collaboration.
Roles—differentiate community members who inhabit specialized niches of performance according to their PKK agency or HPL expertise.	AI **helps members fulfill their community roles** (e.g., publishes, peer reviewers, new member support services, field historians, and theorists)
Rules—includes membership rules, norms, and practices of the global community; includes rules of entry into the community and the rule of entry of new knowledge into the field.	AI systems **remember, maintain, and mirror algorithms as research test beds** for empirical, intuitional, creative, and hypothesis-driven approaches.

1942) of artifacts. Of note here is recent work defining semantic information (e.g. amount of meaningful information) as the syntactic information (e.g. amount of statistical correlations) that a physical system has about its environment that is causally necessary for the system to maintain its existence (Kolchinsky & Wolpert, 2018) and illustrated further by physics-enhanced deep-surrogate approaches towards developing models for complex physical systems (Pestourie et al., 2023). AI can clarify objectives, help set intermediate tasks and monitor progress of not only 'within organization' aims at the meso level, but also between and across organizations that are exploring (or should perhaps consider exploring) partnerships, collaborations, and mergers. AI can alert individuals, teams and intercultural organizations to conflicts and contradictions that raise ethical concerns about the balance of individual agency vs team or organization agency vs cultural agency in a particular case. AI can play a role in the discussion and resolution of questions such as who benefits? Who is at risk of harm? Whose decision is this? Who needs to know? These three areas—Artifact—Object—Subject—align with the original Vygotsky model of artifact as mediator of the objectives of a subject.

Three additional goals for AI follow Engëstrom's additions to Vygotsky: community, roles and rules. In helping to form and evolve new interdisciplinary communities, AI can help expert communities integrate with other fields of knowledge via innovation and collaboration (Xu et al., 2021). AI can assist community role development as outlined at the meso level, for example, it can play roles in publishing and peer reviewing (Hyde et al., 2023), new member and user support services (Huang & Rust, 2021; Li et al., 2021), and fieldwork and theorizing (Christou, 2023).

Finally, AI systems can remember, maintain, and mirror algorithms as research test beds for empirical, intuitional, creative, and hypothesis-driven approaches (Banks, 2010; Behrens et al., 2008; Epstein, 1999; Sokolowski & Banks, 2010).

The following section expands on the roles of AI within the *dynamic CHAT* framework, focusing on the key nodes of artifact, object, subject, community roles, and rules with the addition of irreversible time, dynamics and network theory. As these are suggested aims for the development of AI in education that enhances learning processes, the convention of using verbs such as 'should' and 'would be best if' are intended to reflect the causal and other network connections from micro-to-meso-to-macro as discussed up to this point.

13.3.1 Co-creating Learning Artifacts

- **Narrative Generation:** AI should assist in crafting narratives to explain complex concepts or historical developments, aiding in knowledge dissemination and informing human decision-making.
- **Rationale Generation:** AI should generate and share the reasoning behind its decisions and aim to foster critical thinking and understanding within impacted people and communities.
- **Visualization Tools:** AI-driven visualization tools should not only illustrate abstract concepts but make them interactive as simulations, enhancing comprehension and engagement of complex concepts and relationships.

13.3.2 Enhancing Interdisciplinary Goal Clarity and Planning

- **Objective Alignment:** AI should aid in aligning individual and community goals, ensuring coherence and synergy in collective efforts. When conflicts arise, each agent (individual, team, partnership) should be enabled to decide for itself, with transparent knowledge of the consequences of their decision.
- **Route Optimization:** AI algorithms should suggest optimized pathways or strategies to achieve interorganizational and intercultural community objectives.

13.3.3 Balancing Agency Issues from the Micro and Meso Levels

- **Agent-Aware Learning Support:** AI should provide tailored learning paths and interventions based on agent-level needs at each level. For example, at the micro level via sensing and intervening in the PKK roles and relationships, or at the meso level to help a team achieve its aim, or at the macro level to promote multidisciplinary and intercultural goals.
- **Collaboration Facilitation:** AI should assist forming cohesive multidisciplinary teams and fostering cross-functional collaborations via sensing and promoting HPL roles from meso level that align with the aim of transdisciplinary understanding and the tenets of cultural learning.
- **Knowledge Community Facilitation:** AI should assist each community at the macro level in adding to knowledge, adding members to the community, providing feedback and supporting individual growth of members from novices to experts to enhance community sustainability and improvement.

13.3.4 Facilitating Cross-Cutting Communities and Expert Integration

- **Community Formation:** AI should identify commonalities and facilitate connections between disparate communities, fostering interdisciplinary collaboration.
- **Expert Integration:** AI should aid experts in integrating their knowledge with other fields, fostering innovation and cross-pollination of ideas.

13.3.5 Supporting Community Roles

- **Publication Support:** AI should assist in sharing knowledge, including manuscript preparation, formatting, and peer-review processes and streamlining knowledge and publication workflows.
- **Community Support Services:** AI should support systems for new members, historians, and theorists, aiding in community integration and continuity.

13.3.6 Evolution of Algorithms

- **Algorithmic Research Test Beds:** AI systems should archive and reflect on past algorithms, serving as test beds for various research approaches.

- **Creative Algorithm Generation:** AI should generate innovative algorithms for empirical, intuitive, and hypothesis-driven exploration within multidisciplinary communities.

These directives aimed at the macro level concern how AI can actively contribute to the different nodes within the *dynamic* CHAT framework, aiding in the creation of artifacts, alignment of goals, facilitation of community roles, and the evolution of interdisciplinary knowledge within organizational and cultural learning contexts. It bears saying that the three levels of the complete system model are not separable, because in every culture can be found subgroups with individual members. Instead, the 3-level model is a dynamic network framework for thinking about the interpenetration of the three levels as leverage opportunities for AI to assist with learning processes.

13.4 Limitations and Conclusions

As the aim of the book was to remain focused on learning processes, there are a great many aspects of AI in education that have not been addressed. Broadly, these omitted aspects include the technical aspects of delivering education, the symbolic role of education at local, regional and national levels and the internal and external politics of institutional power which carries the potential for cultural hegemony that are ever-present in large-scale human activities. AI whether defined broadly or narrowly, is a powerful tool for good or ill depending on the hands of humans wielding it. In this book the aim has been to provide a scientific rationale for the agency of learning, not as a preference or social expedience, but as an imperative and therefore a right of life. As a right, several ethical issues arise when conflicting goals of individuals, organizations and cultures clash. The 3-level model asserts that the agency at each level is the guiding principle for ethical decisions within the level, because the imperative of learning adheres to the agency of the level. That is, when considering individual agency, the individual should have all the decision-making power and knowledge needed to make a self-interested decision. When considering a team or organization's agency, then the community structure of the group should have the power (which includes rejecting an individual as a member), and when considering cultural agency, then the most complex process is entailed, and an external standard is needed for survivability and thriving of a shared multicultural future.

Missing from this narrative are the ethics of AI in education when wielded by any other agent than the individual, team or cultural organization for their benefit. It is important to recognize that even with this limitation there are ethical dimensions concerning the potential conflicts among self and others that deserve more discussion. The book *Artificial Intelligence and Learning Futures* for example, spends most of its pages criticizing AI for using the word 'intelligence' with its roots in sociocultural dominance via 'intelligence testing', eugenics and fascistic ideals and lamenting that higher education has become more of a business than a knowledge-producing community (Popenici, 2023). Acknowledging these genuine concerns, the 3-level model is not intended to prevent abuses or craven capitalistic aims, but to provide a principled learning theory framework for enhancing human agency with AI applications.

The 3-level model starts from a definition of intelligence as a natural consequence of an agent's ongoing learning processes driven by the urge to live. That ubiquitous feature of agency—the causal learning cycle—which is the root cause of the drive to expand into the adjacent possible, constantly strives uphill against entropic forces to support the agent's survivability, reproduction and evolving capabilities at individual, social and cultural levels. The learning processes at micro, meso and macro levels ultimately lead to cognition, knowledge and creative capabilities at every level coevolving within the adjacent possible of a local niche in the universe. This book has sought to outline a framework for developing AI applications that can assist learning processes at all three levels.

13.5 Moving Forward

In the concluding chapters the narrative presents open questions raised by critical friends and reviews of the 3-level framework (Chap. 14) and summarizes the model and major conclusions (Chap. 15).

References

Aktas, M. E., Akbas, E., & Fatmaoui, A. E. (2019). Persistence homology of networks: Methods and applications. *Applied Network Science, 4*(1). https://doi.org/10.1007/s41109-019-0179-3

Banks, C. M. (2010). Introduction to modeling and simulation. In *Modeling and simulation fundamentals: Theoretical underpinnings and practical domains* (pp. 1–24). Wiley. https://doi.org/10.1002/9780470590621.ch1

Baum, E. (2003). *What is thought*. MIT.

Behrens, J., Frezzo, D., Mislevy, R., Kroopnick, M., & Wise, D. (2008). Structural, functional, and semiotic symmetries in simulation-based games and assessments. In E. Baker, J. Dickieson, & W. Wufleck (Eds.), *Assessment of problem solving using simulations* (pp. 59–80). Lawrence Erlbaum Associates.

Braitenberg, V. (1984). *Vehicles, experiments in synthetic psychology*. MIT Press.

Castelvecchi, D. (2024). DeepMind AI outdoes human mathematicians on unsolved problem. *Nature, 625*(7993), 12–13. https://doi.org/10.1038/d41586-023-04043-w

Christou, P. A. (2023). The use of artificial intelligence (AI) in qualitative research for theory development. *The Qualitative Report*. https://doi.org/10.46743/2160-3715/2023.6536.

Edelman, G., & Mountcastle, V. (1982). *The mindful brain*. MIT Press. https://mitpress.mit.edu/9780262550079/the-mindful-brain/

Epstein, J. M. (1999). Agent-based computational models and generative social science. *Complexity, 4*(5), 41–60. https://doi.org/10.1002/(SICI)1099-0526(199905/06)4:5<41::AID-CPLX9>3.0.CO;2-F

Feng, S., Hossain, L., & Wigand, R. T. (2017). Network science research dynamic. In *Encyclopedia of social network analysis and mining* (pp. 1–12). https://doi.org/10.1007/978-1-4614-7163-9_110214-1

Huang, M.-H., & Rust, R. T. (2021). Engaged to a robot? The role of AI in service. *Journal of Service Research, 24*(1), 30–41. https://doi.org/10.1177/1094670520902266

Hyde, A., Chodacki, J., & Shannon, P. (2023). *An initial scholarly AI taxonomy* [Other]. Humanities. https://doi.org/10.54900/6p6re-xyj61

Kolchinsky, A., & Wolpert, D. H. (2018). *Semantic information, agency, and nonequilibrium statistical physics*. http://arxiv.org/abs/1806.08053

Langer, S. K. (1942). *Philosophy in a new key; a study in the symbolism of reason rite and art*. Harvard University Press.

Li, M., Yin, D., Qiu, H., & Bai, B. (2021). A systematic review of AI technology-based service encounters: Implications for hospitality and tourism operations. *International Journal of Hospitality Management, 95*, 102930. https://doi.org/10.1016/j.ijhm.2021.102930

Lim, K. W., & Buntine, W. (2016). Bibliographic analysis on research publications using authors, categorical labels and the citation network. *Machine Learning, 103*(2), 185–213. https://doi.org/10.1007/s10994-016-5554-z

Molontay, R., & Nagy, M. (2021). Twenty years of network science: A bibliographic and co-authorship network analysis. In M. Çakırtaş & M. K. Ozdemir (Eds.), *Big data and social media analytics* (pp. 1–24). Springer International Publishing. https://doi.org/10.1007/978-3-030-67044-3_1

Pestourie, R., Mroueh, Y., Rackauckas, C., Das, P., & Johnson, S. G. (2023). Physics-enhanced deep surrogates for partial differential equations. *Nature Machine Intelligence, 5*(12), 1458–1465. https://doi.org/10.1038/s42256-023-00761-y

Pfeifer, R., Bongard, J. C., & Brooks, R. (2007). *How the body shapes the way we think: A new view of intelligence*. The MIT Press.

Popenici, S. (2023). *Artificial intelligence and learning futures: Critical narratives of technology and imagination in higher education*. Routledge.

Sokolowski, J. A., & Banks, C. M. (Eds.). (2010). *Modeling and simulation fundamentals: Theoretical underpinnings and practical domains* (1st ed.). Wiley. https://doi.org/10.1002/9780470590621

Xiong, M. (2022). *Artificial intelligence and causal inference* (1st ed.). CRC Press, Taylor & Francis Group.

Xu, Y., Liu, X., Cao, X., Huang, C., Liu, E., Qian, S., Liu, X., Wu, Y., Dong, F., Qiu, C.-W., Qiu, J., Hua, K., Su, W., Wu, J., Xu, H., Han, Y., Fu, C., Yin, Z., Liu, M., et al. (2021). Artificial intelligence: A powerful paradigm for scientific research. *The Innovation, 2*(4), 100179. https://doi.org/10.1016/j.xinn.2021.100179

Chapter 14
Open Questions

Micro Level issues include:

> *Cognitive Liberty* (Sententia, 2004)—the right to privacy of one's thoughts, self-determination, and individual decision-making, despite transformations in our understanding and ability to manipulate cognitive processes. Cognitive liberty is potentially under threat, for example, due to multimodal sensors and information nets analyzed by AI used for population surveillance in classrooms or research labs using neurotechnology devices to monitor engagement.

Meso Level issues include:

> *AI acting ethically as a self-regulating learner.* Does AI need multiple firewalls to guard against conflicts of interest when acting as an individual and a team member? For example as Sharples notes, is the AI able to "provide verifiable evidence to justify opinions or decisions…[and]… to reason in a way that humans can understand about the "what", "how" and "why" of a continuing conversation" including engaging in constructive conflict of opinions (Sharples, 2023) p.7?

AI acting with empathy towards others. The theory framework outlined here suggested operational roles at the meso level, where empathy towards others is a critical feature of caring about the impacts one might have on others in a team or organizational setting. However, the framework did not directly address the issue of 'theory of mind' (ToM) arising in a generative-AI social learning agent that is simultaneously self-learning while acting as a participant observer on a team.

With ToM, each agent in a multiagent context (e.g. team learning and problem-solving) can:

> "…infer the mental states and intentions of others according to its (local) observation. Based on the inferred states, the agents decide "when" and with "whom" to share their intentions. With the information observed, inferred, and received, the agents decide their

© The Author(s), under exclusive license to Springer Nature Switzerland AG 2024
D. C. Gibson, D. Ifenthaler, *Computational Learning Theories*, Advances in Analytics for Learning and Teaching,
https://doi.org/10.1007/978-3-031-65898-3_14

> sub-goals and reach a consensus among the team. In the end, the low-level executors independently take primitive actions to accomplish the sub-goals. (Wang et al., 2021, p. 1)

Wang and others have demonstrated the idea of ToM in cooperative navigation and multisensor target coverage. Their experiments show that AI with a theory of mind process not only outperforms the state-of-the-art methods on reward and communication efficiency, but also generalizes across different scales of the environment. Their implementation of ToM involves a 2-level agent, which aligns with this model's dual meso level roles of an individual in a private world who is also a team member in a social world. They also use a communication reduction method which may be homologous with 'synaptic pruning' when the multiagent model is housed within a single individual, and is in their case, a public network shared training and execution protocol when the agent is in a multiagent team decision-making setting.

The individual architecture of the process of using ToM in a social team learning context has four components: Observation encoder, Theory of Mind net, Message sender, and Decision maker.

The *Observation encoder* uses an attention-based module, reminiscent of the disequilibration-attention phase of the causal learning cycle in the PKK model. The *Theory of Mind net* enables agents to infer the observation and intentions of other agents, which if inside of an individual would be the subagents of assimilation and metacognition as detailed in Chaps. 4,5, 6, and 7 and the social meso level, those are inferences about the other team members. The *Message sender* acts in three ways to achieve what we've identified in the PKK model as accommodation:

1. It makes an inference about the goals of others (note the similarity to *relevance* in the PKK)
2. It chooses which agent and when to send information (note the similarity to relevance leading to *accommodation* in the PKK)
3. It undergoes a communication redundancy step (which we've likened to synaptic pruning at the micro level and identified as symbolization in the isothermal compression accompanying accommodation)

The *Decision maker* then assigns subtasks to agents (or subagents) who go about their work satisfying the achievement of subgoals towards the team's mission. Note the similarity in the PKK model to confidence building via accommodation and the satisfaction of the task requirements that re-establishes equilibration.

References

Sententia, W. (2004). Neuroethical considerations: Cognitive liberty and converging technologies for improving human cognition. *Annals of the New York Academy of Sciences, 1013*(1), 221–228. https://doi.org/10.1196/annals.1305.014

Sharples, M. (2023). *Towards social generative AI for education: Theory, practices and ethics*. https://doi.org/10.48550/ARXIV.2306.10063

Wang, Y., Zhong, F., Xu, J., & Wang, Y. (2021). *ToM2C: Target-oriented multi-agent communication and cooperation with theory of mind*. https://doi.org/10.48550/ARXIV.2111.09189

SPRINGER NATURE

GPSR Compliance

The European Union's (EU) General Product Safety Regulation (GPSR) is a set of rules that requires consumer products to be safe and our obligations to ensure this.

If you have any concerns about our products, you can contact us on ProductSafety@springernature.com

In case Publisher is established outside the EU, the EU authorized representative is:

Springer Nature Customer Service Center GmbH
Europaplatz 3
69115 Heidelberg, Germany

The manufacturer's authorised representative in the EU is Springer Nature Customer Service Centre GmbH, Europaplatz 3, 69115 Heidelberg, Germany. If you have any concerns regarding our products, please contact ProductSafety@springernature.com

Printed and bound by CPI Group (UK) Ltd, Croydon, CR0 4YY
30/09/2026
02225513-0013